Praise for *Hidden Harmony*

"The book possesses an alluring lyricism and a good sense of humor . . . 'The style of a proof reflects the character of its maker,' declare the Kaplans, and the best part of their book is watching them render editorial judgments on various proofs throughout history . . . One cannot help but think that Pythagoras, who believed in the transmigration of the soul, and who believed his own soul had once belonged to a prostitute, a fisherman, a son of Hermes, and a peacock, among other beings, is more truly reincarnated in all these proofs as they reverberate up through the centuries." —*Boston Globe*

"The Kaplans sinuously weave personalities into the history of proving Pythagoras correct . . . They explore proofs proffered by characters historical (James Garfield), obsessive (Elisha Loomis, a teacher who anthologized proofs), and inspiring (a blind woman, E. Coolidge, who devised an original proof). Born of geometry, the theorem proves to be a creature of metamorphosis, appearing in branches of math from number theory to calculus. Showing the theorem's endless versatility, the Kaplans and their logic- and symbol-permeated text will engage those who delight in doing the math." —**Gilbert Taylor,** *Booklist*

"The Kaplans are wonderfully chatty hosts . . . [They] combine math history and theory with humor, compelling tidbits, and helpful equations (along with an analysis of tangrams) to create an entertaining and stimulating book for the mathematically inclined." —***Publishers Weekly***

"Right up front, [the Kaplans] declare the consensus of historians of mathematics: Pythagoras did not in fact create his eponymous assertion; at most, he might have provided the first compelling proof of its validity. The authors instead trace the roots of the theorem to the dawn of civilization, and to a host of mathematicians, named and anonymous . . . The authors succeed in explaining the arcane aspects of the subject, and they are diligent in situating the Pythagorean theorem within the historical rise of mathematics. That they revel in the subject is clear." —*The Wall Street Journal*

L'arc et la fleche
Mesure du carré de l'hypotenuse
Mont–St–Michel 12th century

HIDDEN HARMONIES

THE LIVES AND TIMES
OF THE
PYTHAGOREAN THEOREM

ROBERT KAPLAN AND
ELLEN KAPLAN

Illustrations by Ellen Kaplan

BLOOMSBURY PRESS
NEW YORK · LONDON · NEW DELHI · SYDNEY

BY THE SAME AUTHORS

The Nothing That Is: A Natural History of Zero

The Art of the Infinite: The Pleasures of Mathematics

Out of the Labyrinth: Setting Mathematics Free

Chances Are . . . Adventures in Probability
(Ellen Kaplan with Michael Kaplan)

Bozo Sapiens: Why to Err Is Human
(Ellen Kaplan with Michael Kaplan)

Published by Bloomsbury Press, New York

All papers used by Bloomsbury Press are natural, recyclable products made
from wood grown in well-managed forests. The manufacturing processes
conform to the environmental regulations of the country of origin.

LIBRARY OF CONGRESS CATALOGING-IN-PUBLICATION DATA

Kaplan, Robert, 1933–
Hidden harmonies : the lives and times of the Pythagorean theorem /
Robert and Ellen Kaplan.—1st U.S. ed.
p. cm.
Includes bibliographical references and index.
ISBN: 978-1-59691-522-0 (alk. paper hardcover)
1. Pythagorean theorem—History.   2. Mathematics—History.
I. Kaplan, Ellen, 1936–   II. Title.
QA460.P8K37 2010
516.22—dc22
2010019959

First published by Bloomsbury Press in 2011
This paperback edition published in 2012

Paperback ISBN: 978-1-60819-398-1

1 3 5 7 9 10 8 6 4 2

Designed by Sara Stemen
Typeset by Westchester Book Group
Printed in the United States of America by Quad/Graphics, Fairfield, Pennsylvania

FOR

BARRY MAZUR

ἔστι γὰρ ὁ φίλος ἄλλος αὐτός

A hidden connection is stronger than an apparent one.

—HERACLITUS

Contents

An Outlook on Insights

*M*an *is the measure of all things.* While Protagoras had the moral dimension in mind, it is also true that we are uncannily good as physical measurers. We judge distances—from throwing a punch to fitting a beam—with easy accuracy, seeing at a glance, for instance, that these two lines

add up to this one:

add up to this one:

What's odd is that we are much less accurate at estimating areas. How do the combined areas of these two squares:

compare to the area of this?

It's hard to believe that they are the same—which is why the Pythagorean Theorem is so startling:

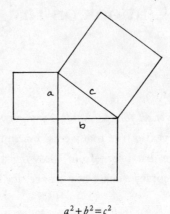

$$a^2 + b^2 = c^2$$

Where did this insight come from—and how do mathematical insights in general surface? How did the tradition evolve of measuring the worth of our insights by proofs, and what does this tell us about people and times far distant from ours—yet with whom we share the measures that matter?

We share with them too the freedom of this shining city of mathematics, looking out where we will from its highest buildings or in astonishment at the imaginative proofs that support them, made by minds for minds. You can safely no more than glance at these proofs in passing, knowing they wait, accessibly, for when you choose to explore their ingenious engineering.

Come and see.

CHAPTER ONE

The Mathematician as Demigod

The Englishman looked down from the balcony of his villa outside Florence. Guido, the peasant's six-year-old son, was scratching something on the paving-stones with a burnt stick. He was inventing a proof of the Pythagorean Theorem; what we remember as the algebraic abstraction $a^2 + b^2 = c^2$ he saw as real squares on the sides of a real triangle—

"'Do just look at this. *Do*.' He coaxed and cajoled. 'It's so beautiful. It's so easy.'" And Guido showed the Englishman's son how the same square could be filled with four copies of a right triangle and the squares on its sides,

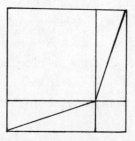

or the same four triangles and the square on the hypotenuse—

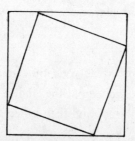

so that the two squares of his first diagram must equal in area the one square in the second.

And the Englishman thought "of the vast differences between human beings. We classify men by the colour of their eyes and hair, and the shape of their skulls. Would it not be more sensible to divide them up into intellectual species? This child, I thought, when he grows up, will be to me, intellectually, what a man is to a dog. And there are other men and women who are, perhaps, almost as dogs to me."

But the child never grew up: he threw himself to his death in despair at being snatched from his family by a well-meaning *signora*, who forced him to practice his scales and took away the Euclid that the kind Englishman had given him.

A true story? Emphatically not. It is Aldous Huxley's "Young Archimedes",[1] published three years after Sir Thomas Heath's *History of Greek Mathematics* came out in 1921, with its quotation from ancient Callimachus: "By a happy chance Bathycles's son found old Thales scraping the ground and drawing the figure discovered by Pythagoras."[2]

The falsity of this story isn't just in Huxley's having patched it together from what he had read in Heath, and from the legend of the brutal centurion who, sent to fetch Archimedes, killed him instead, because he wouldn't stop drawing his diagrams in the dirt. Huxley's story is much more deeply false: false to the way mathematics is actually invented, and false to the universality of mind. The picture of superhumans in our midst—living put-downs to our little pretensions, yet testimony to more things in heaven and earth—is certainly dramatic, but the actual truth has greater drama still, woven as it is of human curiosity, persistence, and ingenuity, with relapses into appeals to the extraterrestrial. How this truth plays out is the story we are going to tell.

△ △ △

A touch of taxes makes the whole world kin. Around 1300 B.C., Ramses II (Herodotus tells us)[3] portioned out the land in equal rectangular plots among his Egyptian subjects, and then levied an annual tax on them.

When each year the flooding of the Nile washed away part of a plot, its owner would apply for a corresponding tax relief, so that surveyors had to be sent down to assess just how much area had been lost. This gave rise to a rough craft of measuring and duplicating areas. Perhaps rules of thumb became laws of thought among the Egyptians themselves; in any event, such problems and solutions seem to have been carried back by early travelers, for "this, in my opinion," Herodotus says, "was the origin of geometry [literally 'land-measure'], which then passed into Greece."

Did their rules, long before Pythagoras, include the theorem we now attribute to him? For Egyptian priests claimed that he, along with Solon, Democritus, Thales, and Plato, had been among their students.[4] Whoever believes that all things great and good should belong to a Golden Age, and that Egypt's was as golden as its sands, would like to think so. But a theorem is an insight shackled by a proof (such as Guido's) so that it won't run away, and not the least line of a proof has been found on Egypt's walls or in its papyri. Well, but might they not have had the unproven Insight, which would have been glory enough?

Here's a shard of what seems like evidence. The Berlin Papyrus 6619,[5] dating from the Middle Kingdom (roughly 2500 to 1800 B.C.), has a problem that amounts to finding the side of a square whose area, along with that of a square with ¾ its side-length, sums to a square of area 100. Since the solution to what we would write as $x^2 + \left(\dfrac{3}{4}x\right)^2 = 100$ is $x = 8$, giving squares of sides 6, 8, and 10; and since these three numbers are doubles of 3, 4, 5—the most basic, and famous, triple of whole numbers that fulfill the Pythagorean relation—

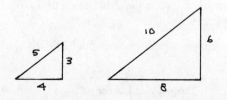

this could be muffled testimony to knowledge of how the squares on the sides of at least a particular right triangle were related.* Of course no triangle appears in this problem, much less a right triangle; but if you hear a faint "Yes" when you ask "Is anyone there?", it would be natural to assume that someone indeed was.

Or it might be that 3, 4, and 5 came up just through learning your times tables. After you found out that $3 \times 3 = 9$, and then that $4 \times 4 = 16$, you might well have been struck, on discovering that $5 \times 5 = 25$, that this was the sum of the two so recently learned (our pattern-making instinct seizes on such slantwise connections). A discovery like this might fall into the background of thought when the pattern failed to continue ($4^2 + 5^2 \neq 6^2$) and surface again when constructing framing for doors—or accessible problems for students. To disguise what might have been the too obvious (3, 4, 5) as (6, 8, 10) would have been a teacher's familiar dodge. But if the only Pythagorean triples we find among the Egyptians are multiples of (3, 4, 5), the likelihood grows that this generator was for them no instance of a broader truth, but just an attractive curiosity.

Here, however, is a second shard—needing more imagination to decipher than the first, and therefore more intriguing. Early in the fifth century B.C., the philosopher Democritus bragged that none had surpassed him in geometric constructions and proofs, "not even the rope stretchers of Egypt"—apparently referring to the surveyors who, from as early as 2300 B.C., used ropes and pegs to lay out the corners of sacred precincts. Long after this, ancient Indian and perhaps Chinese geometers made right angles by stretching ropes knotted in lengths of 3, 4, and 5 (so that a triangle made with them would have to have been a right triangle). By thus stretching imagination itself back to the time of Amenemhat I, we might feel convinced that the full Pythagorean Theorem, and not just this mingy triple, had been a permanent feature of the Egyptian landscape. You could go on to let living tradition testify to its antiquity: a stonemason of our acquaintance blithely uses what he calls

---

* A similar problem in this papyrus has the solution (12, 16, 20), which is the triple (3, 4, 5) multiplied by 4.

the 3-4-5 rule for checking that his foundations are square. And is not his craft directly descended from the Egyptians, Indians, and Chinese?*

Fantasy more than imagination is needed to interpret a third offering of evidence, stretching this rope still farther. The engineer and archeological theorist Alexander Thom surveyed, in the middle of the last century, hundreds of megalithic standing stones in northern Europe, and concluded that a standardized 'megalithic yard', and the properties of Pythagorean triangles, had been used in their placing. Followers spread his conclusions southward, so that, as one of them writes, "It should therefore perhaps not come as a surprise that in 1997, a stone circle was found in Egypt, in Nabta, which was dated to roughly the same period as the stones at Carnac. It introduced a megalithic dimension in Egypt. At the same time, the Nabta circle had clear astronomical components." Unfortunately, a statistical analysis of Thom's data for 465 possible triangles led to the conclusion that "the ancient megalithic builders did not understand the properties of Pythagorean triangles . . . and the idea that they might have is really an artifact of modern interpretation, reasoned a posteriori according to our modern-day knowledge of Pythagoras' theorem."[6]

To whatever slight extent the Egyptians had the Pythagorean Insight, where did it come from? Or if their surveyors had neither knotted ropes nor megalithic yards in their backpacks, how did the Greeks then manage to come by this knowledge? A desert of ignorance drifts for us over these vast centuries; yet turn northeast from Egypt and backward in time, and look across the sands to Mesopotamia . . .

---

* George Sarton, *A History of Science: Ancient Science Through the Golden Age of Greece* (Cambridge: Harvard University Press, 1952, 39), thinks that the Pythagorean interpretation of rope-stretching is a red herring, being a misunderstanding of an astronomically-slanted ceremony ("Stretching the cord") at the founding of a temple.

CHAPTER TWO

Desert Virtuosi

The deeper in time the Golden Age is set, the more romantically it gleams. Some four thousand years ago, between the Tigris and Euphrates rivers, the Old Akkadians and then the Old Babylonians developed a way of life that bustled with pride and commerce. They did things with numbers and shapes of a finesse and intricacy that will take your mind's breath away. These people were the contemporaries of your great-great-[some 150 of these] . . . great-grandparents; they stood two-thirds as tall and lived half as long as we; had a hundredth of our comforts and none of our safeguards. They had no Twitter, no dentists, no Big Macs—but their sense of humor puts them just down the street from us.

> FATHER: Where did you go?
> SON: Nowhere.
> FATHER: Then why are you late?

This snatch of dialogue has been deciphered from an ancient cuneiform tablet, startlingly small, and indented with the neat bird-tracks of wedges that read as easily to them as our letters do to us.[1]

The ancestors of these people had kept their accounts with clay tokens through the four thousand preceding years, but as a temple-based bureaucracy developed, the growing complexity of life, and of the bookkeeping that recorded it, led to symbols for these tokens, and signs for 1, 10, and 60, which they iterated to make the other numbers. Once again we recognize ourselves in them: abstraction from things to names, and names to numbers, is the way we mark our turf.

An organic interplay between mathematics and administration

continued through most of the third millennium B.C., developing into a 60-based system of whole numbers and fractions (60 has enough whole number divisors to make calculations easier than in our base 10 system). Then something important seems to have happened around 2600, when a class of scribes emerged. For them—perhaps in the slack periods when goods were not being brought in—writing broadened from making inventories to recording epics, hymns, and proverbs, and mathematics from the practical to the precious. We set our young mathematicians difficult theorems to prove; they gave their rising scribes horrendously long calculations to carry out, with as little connection to reality as had the highly artificed poems that Mandarin officials were required to write in ancient China.

Everything changed again near 2300, with the invasion of an Akkadian-speaking dynasty. Sumerian stiffened into an administrative language only (playing the ennobling role that Latin once did for us), and new sorts of mathematical problems arose, centering on area. In the blink of an eye from our perspective—two hundred years for them—this dynasty fell, and a neo-Sumerian state took its place in 2112. That 60-base number system now began to work itself out on the backs of the people and the brains of the scribes. Our lawyers punch in a client's time every fifteen minutes, but these scribes, acting now as overseers, had to keep track of their laborers through a day of ten-minute quotas.

This state in turn collapsed—probably under its own administrative weight—within a century, and the four-hundred-year glory of the Old Babylonian period began, epitomized by the famous lawgiver Hammurabi. This was a time of high scribal culture, featuring ideals we recognize as humanistic, and calculations whose balance of cleverness and painstaking tedium we gasp at.

A Hittite raid around 1600, then an overwhelming invasion by warrior Kassites, suddenly brought down a curtain a thousand years thick, hiding away almost all traces of this Hobbit-like people, whose glass-bead game culture and animated bureaucracy we see far away down the wrong end of a telescope.[2] Our story turns into history.

Both are deceptive. This narrative, made to appear seamless, is actually stitched together from so much of so little—as far as the mathematics goes, mountains of clay tablets recording hardly more than school exercises or teachers' trots. The context of the society at large, and of the scribal community within it, is just guesswork, attempts at rational reconstruction baffled by the distortions of historical foreshortening: events spread over vast stretches of time and space are collapsed to aperçus, and anything is taken to stand for everything. The conclusion, for example, that a thuggish regime shut down intellectual pursuits for a millennium might be skewed by the economics of excavation: with funding scarce, who would dig up schoolrooms, despite their possibly valuable evidence of evolving thought, when there are palaces waiting to emerge?

We're not only sitting at the far end in a game some call Chinese Whispers and others Telephone, but those in front of us each have their own agendas and personalities to promote, while they indulge in the peculiar practice of letting their civility be seen as no more than veneer. Perhaps they would be less vituperative, and hence more enlightening, were they a more expansive community (squabbling seems to breed in close quarters), or had they evidence rather than speculation to go on, and a logic founded on 'only if' rather than 'if only'. As it is, the hum of knives being whetted may serve the ends of mean fun, but can be distracting. From a recent scholarly work:

> The pretentious and polemical attempt by Robson in *HM* 28 (2001) to find an alternative explanation of the table on Plimpton 322 is so confused and misleading that it should be completely disregarded, with the exception of the improved reading of the word *I-il-lu-ú* in the second line of the heading over the first preserved column, and the dating of the text. . . . Cf the verdict of Muroi, HSJ 12 (2003), note 4: "The reader should carefully read this paper written in a non-scientific style, because there are some inaccurate descriptions of Babylonian mathematics and several mistakes in Figure 1, Tables, and transliterations." A briefer and less polemical, but still pointless, version of the same story can be found in Robson, AMM 109 (2002).[3]

Odd that archeology is often still thought of as one of the humanities.

Our aim at this point is to see what traces or precursors of the Pythagorean Theorem we can find in Mesopotamia; and if there are any, whether they then migrated somehow to Egypt or even directly to Greece. We recognize that the ambiguity of the evidence and the need to pick at the scholarly scabs over it will make our telling hypermodern: one of those "design it yourself" dramas, constructed not only by the participants but by its writers and readers as well. Yet some Guido or, collegially, Guidos did come up with the stages of this insight, and then its proof; and narrowing down to a local habitation if not a name is surely less profound a task than was theirs.

△ △ △

From what was practiced in their schools, the Old Babylonians appear to have prized virtuosity over curiosity, technique over insight. Whether they were simply addicting students to the calculations they would need as administrators, or whether computation induced an almost erotic ecstasy (as anything chant-like and repetitive may), we find countless examples of methodological rather than critical thought, of assertions rather than explanations. Theirs was, at first glance, a culture of algorithm. Hammurabi's Code of Laws well represents their point of view: it is a collection of particular judgments that serve as precedents for subsequent decisions. Each of their mathematical problems too is an example, none is exemplary: every figure has particular numbers attached to it, such as lengths and areas: there is no "general figure" (while a triangle in Euclid's geometry stands for any triangle of its ilk) and certainly no notion of a "general number"—the x that, millennia later, would lift arithmetic to algebra.

What kinds of problems were these? Additions and subtractions, multiplications, difficult divisions (dividing sexagesimal—base 60—numbers by anything but multiples of 2, 3, and 5 leads you into fearful mazes of remainders). They also had a curious predilection for questions of this sort: find a number which, minus its reciprocal, equals 7. This fascination with reciprocals is explained by their thinking of division—by 5, say—as multiplication by the reciprocal of 5 (which was, for them, not ⅕, but 60/5).

They calculated everything in sight: when given some dimensions of, say, a door—as its width and length, to find its area, or the length of its cross-brace. Ah! You mean to find the hypotenuse of this right triangle?

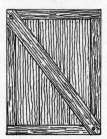

So it seems—as in their problems about canes leaning against walls: how long are they, if their tops are this far from the ground, and their tips that far from the wall?

So they must have had the Pythagorean Theorem! Well, let's look more closely.

How were they taught to solve these problems? By reckoning, not reasoning. They were given procedures: if it looks like *this*, do *that* to it (perhaps your math teacher was an Old Babylonian?). Even the instructors worked from handed-down tables of algorithms and answers. Yes, but somewhere, somewhen, someone had come up with the insight that led to these algorithms. Of course—and these algorithms are their only memorials, as well as memorials to a time, perhaps, when curiosity was commoner coin. But we need to understand the context of these algorithms, if we are to know whether any of them incorporated the Pythagorean Insight.

We mentioned that the mathematical issues the Old Babylonians dealt with had come to center on area. You now begin to sense, after the first shock of recognition across the millennia, how very different their mathematical intuition was from ours. We build up square measures from linear, areas from lengths. For some reason their thought moved in the contrary direction—so much so, that they used the same word, *mithartum*,[4] for a square and its side. Absurd? As absurd as your asking that a square and its side and the number −2 add up to 0? An ancient Greek would surely have pointed out the nearest lunatic asylum[5] to you: you can't add areas to lengths, much less either to a dimensionless number (and a negative one at that). Yet we blithely write $x^2 + x - 2 = 0$ and, solving, find $x = 1$ or $x = -2$: pleased not only that the square has a side of length 1, but that this side could equally well have had length −2. What is a square, and what is its side, and unto whom?

But when we are thinking geometrically, rather than algebraically, we, along with the Greeks, would still insist that lengths and areas are fundamentally different. For what may, as we'll now see, have been ingenious methodological ends, the Old Babylonians blurred this distinction: so much so, that they often "thickened" lines to regions constructed on them. Since prohibitions signal the presence of the practice they legislate against (as Puritan strictures against colorful clothing tell us of the russets, ochres and olives the settlers wore), does an echo of this Babylonian eccentricity explain why, long after, Euclid goes out of his way to *define* a line, and in his definition says that "a line is *breadthless* length"?

To see how the Old Babylonians benefited from what for us would be confusion, and to see the bearing it has on the Pythagorean Insight, look at how they dealt with problems of what they called *igi* and *igi–bi*: numbers and their 'reciprocals'. Were we asked to find a number which, minus its reciprocal, equals 7, we would let x stand for the number, so that its reciprocal would be $1/x$, and try to solve for x with $x - \dfrac{1}{x} = 7$. Remember, though, that we're supposed to be role-playing Babylonians, and that their sexagesimal system meant that they worked in terms of 60 as a unit (as we do when thinking of 60 minutes in an hour). They were therefore aiming for a pair of *whole* numbers, x and $60/x$ (these would be reciprocal, since their product would be 60), whose difference was 7. It tells you something about them that where we write 'product', they said "Make x and $60/x$ eat each other."[6]

What we need, then, is to solve for x when $x - \dfrac{60}{x} = 7$. We would blithely multiply both sides by x, rearrange to get $x^2 - 7x - 60 = 0$ and factor (or use the quadratic formula) to find $x = 12$ and its reciprocal 5: and indeed 5 times 12 is 60.

Not so easy for the real Old Babylonians. Thirty-five hundred years ago they not only had no symbol for the unknown, they had no quadratic formula, no algebra, and no equations. They had, however, lines with breadth. So draw, as they did, a square box, and label (for our convenience—this would have been alien to them) both its height and width $60/x$.

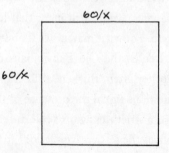

Now *thicken* the right-hand line by $x - \dfrac{60}{x}$ —i.e., by 7 (it's probably easier for us simply to think of pulling the right-hand edge to the right), so that we now have a rectangle $\dfrac{60}{x} + x - \dfrac{60}{x} = x$ wide and, like the square it was thickened or pulled from, $60/x$ high; and our whole new figure is a rectangle $60/x$ high and x wide—whose area is therefore 60:

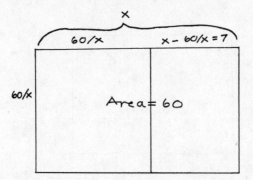

They now changed their visual metaphor from thickening lines to tearing areas (did they make their mathematical manipulatives out of clay?). They took this rectangle with area 60 and tore the new, right-hand part of it vertically in half, making two rectangles each $7/2$ wide but still $60/x$ high:

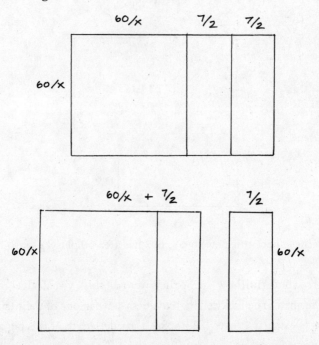

Take, with them, this detached, narrow right-hand rectangle, give it a quarter turn, and hang it (as they wrote) underneath the original square:

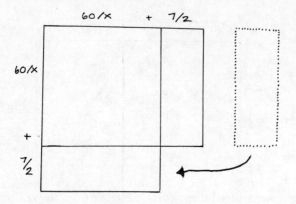

This gave them an indented figure with area 60 (it's just the parts of the x by $60/x$ rectangle rearranged). Notice that the missing piece of this figure is a $7/2$ by $7/2$ square: a square whose area is therefore $49/4$, or $12\frac{1}{4}$.

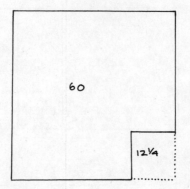

Altogether, then, this new square box has area 60 plus $12\frac{1}{4}$, so that its total area is $72\frac{1}{4}$.

It's wonderful how adept they were at this "visual algebra". But now they had to be lucky enough to guess, have in one of their tables, or be able to calculate, that $\sqrt{72\frac{1}{4}} = 8\frac{1}{2}$, the side-length of this big square. That means

$$\frac{60}{x} + \frac{7}{2} = 8\,\tfrac{1}{2},$$

so $\dfrac{60}{x} = 5$ and $x = 12$: the number x for them, as it was for us, is 12, and its reciprocal, $^{60}/_x$, is 5.

Magical! What an ingenious way to solve this problem (for all that the Babylonian teacher had designed it to give a nice answer). But something yet more magical happens if you take a step back from the engineering we've just been immersed in. That square, with its little square tucked into a corner—where have we seen it before?

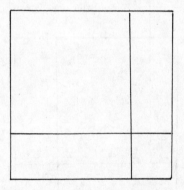

Drawn by Guido in the dust of Chapter One! It lacks, of course, the crucial diagonals in the two rectangles that let us form and move triangles around to prove the Pythagorean Theorem—but it is the nurturing *context* for that insight.

In fact this diagram—let's call it the Babylonian Box—seems to have been the key feature, for a thousand years, of how these people did their area-centered math—as well it might be: for looked at one way, it represents (as we would say) $(a+b)^2$:

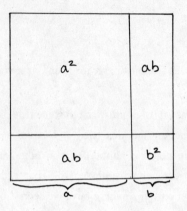

Looked at in another and rather subtler way, it can represent $(a-b)^2$:

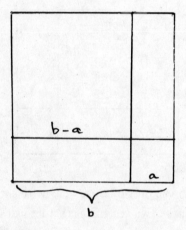

If, in a box with side b, a box with side a is tucked into the corner, the remaining square of side $(b-a)$ has area b^2 less the two ab rectangles (minus their a^2 overlap):

$$(b-a)^2 = b^2 - (2ab - a^2) = b^2 - 2ab + a^2$$

Pretty much anything that we can do with second degree equations, they could do via their box. This doing may lack the great leaps that abstraction allows, but it wins in agility: *tearing off* part of a rectangle, *turning* it, *hanging* it under another. Such liveliness goes some way toward making up for the mind-numbing calculations they devoted themselves to. It also reminds us, once more, how very different their way of being human was from ours. Hours, days, lifetimes spent in dark computations unlit by wonder, and the same old box of a diagram trotted out endlessly—and then suddenly, a variation on that slender theme opening onto a new world.

Here is that box:

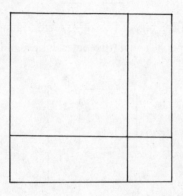

Now *tear* its backward L away and *hang* it a little bit down from the remaining square:

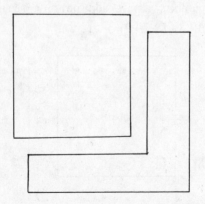

That L (just the right and bottom edges thickened) is what the Greeks, long after, called a gnomon, which we still use decoratively in our gardens to tell sunny time by the shadow it casts on a dial. It was introduced into Greece from the Babylonians, says Herodotus—perhaps by the philosopher Anaximander.[7] And look at what you can do if you pave it with pebbles! Put one in the box left behind, then three in the surrounding gnomon: since the square plus the gnomon always forms a new square, our $1 + 3 = 4$.

Do this again with the bigger box of 4, surrounding it with a gnomon that must have five pebbles in it (two on each side, and one in the corner): $1 + 3 + 5 = 9$.

If you keep doing this, you'll see that since the gnomon always has an odd number of pebbles in it ($2a + 1$, when the inner box has a^2), the successive odd numbers always add up to a square: $1 + 3 + 5 = 3^2$, $1 + 3 + 5 + 7 = 4^2$, $1 + 3 + 5 + 7 + 9 = 5^2 \dots$

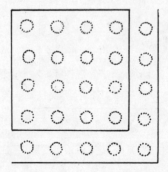

Every square number is the sum of such a sequence of odds, and every sequence of odds, from 1 on, adds up to a square: this glimpse into the profundities of how the natural numbers behave, long attributed to the Pythagoreans,[8] comes from nothing more than a Babylonian Box filled with pebbles, and split into a square and its gnomon. And was this in turn the ancestor of the Greek counting board, and later, the abacus?

But our focus is on the Pythagorean Insight, and how this bears on it. Notice that when the odd number of pebbles in the gnomon is *itself* a perfect square, as $9 = 3^2$, our diagram shows us—*without a triangle in sight*—that $3^2 + 4^2 = 5^2$: that simplest of all Pythagorean triples (which, to be faithful to the diagram, we might better write as $4^2 + 3^2 = 5^2$). In our symbolic language, the big square, $(a+1)^2$, is the smaller square a^2,

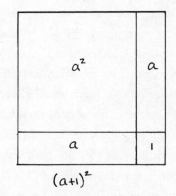

along with the gnomon $2a+1$, and we're looking at the instances when $2a+1$ is a square. Should the triple 3, 4, 5 (or multiples of it) therefore come up in Old Babylonian mathematics, we'll be neither surprised nor tempted to think of it as evidence for the Insight. They might have stumbled on it as well in the way suggested in Chapter One (adjacent entries in times tables), with both of these encounters then reinforced by the importance of 3, 4, and 5 as divisors of 60: this trio was a prominent part of that limited repertoire of "nice" divisors.

In fact, (3, 4, 5) and some multiples of it *do* occur in fifteen of the many problems scholars have deciphered so far; (5, 12, 13) occurs twice, (7, 24, 25) three times, and (11, 60, 61) once. Each of these fills the Pythagorean prescription, and each comes from the fitting of a pebbled gnomon to a pebbled square, as you've seen:

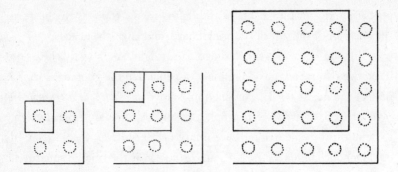

Let's call these 'gnomonic triples'. Of course, there will be an infinite number of these—one for each time the odd number of pebbles in the gnomon is a perfect square (9, 25, 49, 81, 121, . . .). While no others in this sequence have been found on excavated clay tablets, we can picture the Old Babylonians seeing that there would always be a next, world without end. This gives the giddy sort of sensation that often leads people into mathematics: grasping something infinite via abstraction (as children love dinosaurs, because they are both very big and not quite real). But the Old Babylonians may have been too attached to this particular case and then that, ever to have thus broken free.

If we indulge, however, in a brief fantasy, we could imagine that one or another of them saw this infinite series of gnomonic triples, and thought: there they are, all of them, each an instance of two squares adding up to a third—and every time two squares add to a third, they will do so in this way. We could excuse them for leaping to this last conclusion: if you see endlessly many, how could there be others that didn't stand in a row with them? But suppose our Old Babylonian had an afterthought: we got these by fitting a gnomon around a square. Couldn't we balance this action by fitting a second gnomon around the other two sides? Well, probably we'll just be skipping through the sequence we had before—let's see:

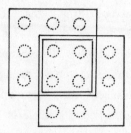

No, this needs *two* more pebbles in it, at the empty corners, to make the big square:

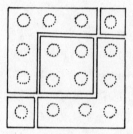

Then this ring of $4a+4$ dots, when it surrounds the inner square of a^2 dots, will be a doubly gnomonic triple! When will this happen? Only when $4a+4$ is a square number—as, for example, when $a=15$, for then $4a+4$ is $64=8^2$, and we get the triple (8, 15, 17)—which wasn't in our other series at all! Looking at this from a modern, algebraic standpoint, however, removes both the mystery and its power: $(a+2)^2 = a^2+4a+4$.

This may be our self-indulgent reconstruction, but in fact the triple (8, 15, 17) *does* come up, twice, on their tablets, though no others of this second sort do. It completes the collection of all the Pythagorean triples found on the thousands of Old Babylonian tablets,[9] and would have been only the beginning of wonder, for them, at the infinite variety of things. None of these triplets, as you see, depends on the Pythagorean Insight; all follow naturally from gnomonic play with their magical box.

△ △ △

"It completes the collection of all the Pythagorean triples . . ." unless—

Unless we fall under the spell of Otto Neugebauer, and The Curious Case of the Babylonian Shard. Neugebauer was the great force in mid-twentieth-century studies of ancient scientific thought. His erudition, acumen, and touch placed him first among his peers. In 1945 he published a broken tablet, dated to about 1760 B.C., illegally excavated in the 'twenties, bought from a dodgy dealer (who had listed its contents as "commercial account") by the publisher George Arthur Plimpton, and later bequeathed to Columbia University. There it still rests, as Plimpton 322. Neugebauer quickly saw that it was nothing like a ledger book, but something quite extraordinary indeed. From what heaven of invention had this slab dropped?

He recognized that its fifteen rows of numbers, in the four extant columns, were mathematically sophisticated, and it struck him that they were two of the three parts of the most astonishing Pythagorean triples. Only one was gnomonic; the rest were gigantic trios that it would have taken the skill of the most sophisticated Greek mathematician, fifteen hundred years later, to come up with*: (119, 120, 169), for example, and (4601, 4800, 6649). And what did the Babylonian scribes, what would anyone, want with such huge numbers—had we mentioned (12,709, 13,500, 18,541)? Deduction, inspiration, and a generous analogy to modern practice (what we can do, so could they, since mathematics belongs not to Culture but to Mind) went into Neugebauer's reconstruction, and into his conclusion that what he took to be the tablet's collection of Pythagorean triples was made for its own sake, in the cause of number theory—the purest of abstract mathematics. At the end he touchingly wrote:

> In the "Cloisters" of the Metropolitan Museum in New York there hangs a magnificent tapestry which tells the tale of the Unicorn. At the end we see the miraculous animal captured, gracefully resigned to his fate, standing in an enclosure surrounded by a neat little fence.

---

* You will see in Chapter Seven this beautiful play of untrammeled thought.

This picture may serve as a simile for what we have attempted here. We have artfully erected from small bits of evidence the fence inside which we hope to have enclosed what may appear as a possible living creature. Reality, however, may be vastly different from the product of our imagination: perhaps it is vain to hope for anything more than a picture which is pleasing to the constructive mind, when we try to restore the past.

Enter The Irascible Scholar and The Invisible Man. Evert Bruins— according to a current writer—laced his papers on Babylonian mathematics with venomous hyperbole, and had what is delicately described as "extraordinarily difficult personal relationships with other scholars."[10] These apparently included charging Neugebauer with sneaking into the Plimpton collection and "trying to break off a piece from Plimpton 322 in order to make the counter-evidence to his theory disappear."[11] The counter-evidence was Bruins', published in a journal he edited and supported financially. It backed up his claim that there wasn't a Pythagorean triple anywhere in Plimpton 322. Although his scholarship was monstrously careless, his style verging on the incomprehensible, and his generalizations ridiculously broad,[12] his interpretations were honored (as another scholar puts it) thanks "to the general conviction that nobody advances devastating criticism without support in strong arguments or indisputable facts."[13]

We have to look at the tablet (with its scribal errors corrected by contemporary scholars) in order to understand the opposing views.

What's at stake for us is the antiquity of the Pythagorean Insight: if these entries are indeed Pythagorean, they could have come in no way (only one being gnomonic) but from a deep knowledge of the Theorem, which would put its origin back at least to 1760 B.C.

(1:)59:00:15	1:59	2:49	1
(1:)56:56:58:14:50:06:15	56:07	1:20:25	2
(1:)55:07:41:15:33:45	1:16:41	1:50:49	3
(1:)53:10:29:32:52:16	3:31:49	5:09:01	4
(1:)48:54:01:40	1:05	1:37	5
(1:)47:06:41:40	5:19	8:01	6
(1:)43:11:56:28:26:40	38:11	59:01	7
(1:)41:33:45:14:03:45	13:19	20:49	8
(1:)38:33:36:36	8:01	12:49	9
(1:)35:10:02:28:27:24:26	1:22:41	2:16:01	10
(1:)33:45	45	1:15	11
(1:)29:21:54:02:15	27:59	48:49	12
(1:)27:00:03:45	2:41	4:49	13
(1:)25:48:51:35:06:40	29:31	53:49	14
(1:)23:13:46:40	56	1:46	15

*Plimpton 322 and its transliteration (additional
columns may have been broken off segment to the left,
the nearest of which may have contained,
in each row, the equivalent of 1).*

The first of its columns, reading with the Babylonians from right to
left, contains the numbering, 1 to 15, of the rows. The second and third
(let's call them columns II and III) are the numbers in dispute: the

short side and the hypotenuse (if Neugebauer is right) of a Pythagorean triple—the long side was presumably in one of the columns on the tablet's missing third. Column IV has sizable numbers in it, differently understood by each of the now several parties to the argument.

Were Neugebauer's view correct, why would the third member of the triple be separated from the other two by an intervening column—and what role, in fact, does that strange column play? Why would these or similar vast triples not show up elsewhere on their tablets? Why were just these fifteen triples chosen, seemingly at random, out of some hundred possible Pythagorean triples that could be made with divisors of 60?* More generally, how would so abstract an end as this suit with the Old Babylonians' notoriously algorithmic and example-bound practice? And why, if they were so keen on Pythagorean triples, has not a single cuneiform tablet been dug up bearing a right triangle decorated with squares?

One of Neugebauer's followers[14] suggested that the obscure fourth column recorded squared cosines or tangents of the angle opposite the short side, and that the rows decrease roughly one degree at a time. This would be remarkable, since the Old Babylonians had little concept of angles other than right angles (even these showed a fair amount of wobble), and none at all of measuring them, much less calculating their cosines or tangents.[15]

Which brings us back to The Irascible Scholar—but in a roundabout way. In 1980 an urbane mathematician named R. Creighton Buck wrote a brilliant paper, "Sherlock Holmes in Babylon",[16] in which he looked again at Plimpton 322 and at Bruins' work (perhaps some of the bile had dissipated by then). He cites the proposals of a D. L. Voils, which (you find in his references) were soon to appear in *Historia Mathematica*. In the nearly

---

* 159, according to Robson, "Neither Sherlock Holmes nor Babylon" 177–78, reduced from the $946 = \dfrac{(44 \times 43)}{2}$ possible pairs of generators available in the 44 numbers of a standard table of reciprocals, if we assume that this already astounding scribe was also familiar with the idea of numbers being relatively prime.

three decades since then, no paper by Voils has surfaced, there or else-where. He is unknown to Wikipedia. The Web produced a phone num-ber, which we called. Whoever answered told us: "There is no Mr. Voils." Thus had we heard, and did in part believe. He is The Invisible Man.

Invisible Men write Disappearing Documents. The ghost of D. L. Voils haunted our days, until at last diligence tracked him down in Florida. He had written the paper, he said, which was for some reason rejected, and he chose not to rewrite it. When he moved from Wiscon-sin he left it behind with other papers, which his son subsequently sold. Since matter is neither created nor destroyed, only changed in form, the atoms of his argument wait somewhere to be reassembled.

What Buck says that Voils suggests, based on Bruins, is that Plimpton 322 was a teacher's trot for making up igi-igi-bi problems!* These were, as you saw, a stock-in-trade of the schools, since they taught the solution of what we call quadratics by manipulating the Babylonian Box. Some were of the form $x + \dfrac{60}{x} = c$, a constant, some (like the one you saw above, pp. 12–15), $x - \dfrac{60}{x} = c$. While tables of reciprocals were a commonplace, what a teacher needed to know was which c yielded nice solutions—and here was the answer.

Between the three of them (we can say 'between', because Voils is the invisible middle term), this is their explanation. Column II involves the differences the teacher would like to set his problem equal to (like the 7 in the illustration we gave on p. 12), column III bears on the sums, and IV serves to check an intermediate step. But where did they get their val-ues from?

Take, our trio says, any numbers you like, a and b (perhaps listed in columns on the missing third of the tablet), and set $\dfrac{a}{b} = x$, the igi, so that

---

* An error in it, easily explained as a mistake in copying, and another record-ing the square of a number rather than the number itself, suggest to some that this was a copy, which in turn suggests a teacher's duplicate from a master.

$^{b}/_{a}$ will be its reciprocal, igi-bi (let's write it x^R, with R for 'reciprocal', so that we won't be wedded to some particular power of 60 in the numerator). Then

$$\frac{a}{b} - \frac{b}{a} = \frac{a^2 - b^2}{ab},$$

and that difference in the numerator, $a^2 - b^2$, is what is listed in column II.

To see how it works, take the entries in row 1: 119 is in the difference column, II, and if we play around a bit, we'll find that a could be 12 and b, 5: for then $a^2 = 144$ and $b^2 = 25$, so $a^2 - b^2$ is indeed 119. The teacher can now set up his problem: find x and x^R if $x - x^R = 7$ (since this is $12 - 5$). He knows the answers will be $^{12}/_5$ and $^5/_{12}$—or, since he doesn't deal in such fractions, he expects his students to proceed in the same way we showed above (pp. 12–15) with $x^R = \dfrac{60}{x}$, not $^1/_x$, and come up with 12 and 5 as the solution—for in fact, this is the very same problem! He could also have made up the problem $x + x^R = 7$, since $12^2 + 5^2 = 169$, as he finds in column III. Notice that the a and b and the sums and differences of their squares are his secret knowledge: the student knows only the cleaned-up statement of the problem, and the instructions for using the diagram:

> The igi-bi over the igi is 7 beyond
> The igi and the igi-bi are what?
> You:
> The 7 that the igi-bi over the igi is beyond, to two you break, then 3½
> 3½ with 3½ let them eat each other, then 12¼
> To 12¼ that came up for you, 60, the field,
> add then 60 and 12¼
> The equalside of 60 12¼ is what? 8½
> 8½ and 8½ its equal lay down
> Then 3½ the holder

> From one tear out
> To one add
> One is 12 [8 1/2 + 3 1/2]
> The second 5 [8 1/2 − 3 1/2]
> 12 is the igi-bi, 5 is the igi.[17]

Be glad you went to school three millennia later.

If you wonder at the stubborn butting of heads against calculations by hand, in cuneiform, with a base 60 system, encumbered by awkward notation, of such quantities—so do we. Conceivably Plimpton 322, with its large values (although 12,709 in row 4 of column II, for example, *only* needs *a* to be 125 and *b*, 54), marks a transition from a geometric to a more formulaic style. But for all that squares and their sums and differences are in play, of Pythagorean triples we haven't a shadow cast by the slimmest of gnomons—and that, Watson, is what we wanted to know.

Since Bruins is undeniably behind this explanation, it's as if someone with disgusting habits were punching you in the stomach while screaming the truth in your ear. You'd hate to admit he was right. But then, we do have him at two removes, and Buck's manner is charming. Eleanor Robson, his severest critic, takes him to task (in "Neither Sherlock Holmes nor Babylon")[18] for being a gentleman-amateur and therefore trying to show up the plodding professionals; for speaking of Babylon at large, when the tablet really came from the city of Larsa; for being (somehow) part of the colonization, appropriation, and domestication of the pre-Islamic Middle East by Western society; and for misinterpreting Voils' theory—which she knows of only through him. She concludes by agreeing with his interpretation. The criticism of her paper by a fellow scholar you have already seen (p. 8).

△ △ △

Neugebauer's unicorn has turned out in the end to be indeed a fabulous beast. A slab delivered to earth by superior beings might have had such triples on it, but not a bit of clay worked by Babylonians. There is, however, one more thread to follow that might yet lead to the Pythagorean Insight in ancient Mesopotamia. Rather than a slab, this is a tablet the size of your palm, numbered YBC 7289, which sits snugly in Yale's Babylonian Collection—and on it is incised a square scored with two diagonals into four congruent isosceles right triangles. Across one of these diagonals is written, in cuneiform, "1 24 51 10"—i.e., $1 + \frac{24}{60} + \frac{51}{3600} + \frac{10}{216000}$: which translated into our base 10 is 1.414212962—an approximation correct to five decimal places for $\sqrt{2}$. Here is that tablet, full size:

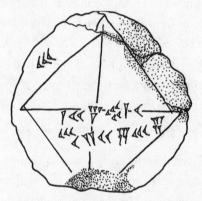

Quite apart from the question of how they managed to calculate $\sqrt{2}$ so precisely, a right triangle with sides 1, 1, $\sqrt{2}$ is certainly not gnomonic. Mustn't they have had the Pythagorean Insight after all in order to know that the diagonal had length $\sqrt{1^2 + 1^2} = \sqrt{2}$?

The answer lies twelve hundred miles west and thirteen hundred years in the future from the Old Babylonians. This diagram, differently scaled, plays a key role in the dialogue *Meno*, which Plato wrote in fifth-century B.C. Athens. In it, Socrates is shown eliciting from a slave-boy what the *mithartum* (as those Babylonians would have said)—what the side-length—must be of a square whose area is 8. Implicit in the text is this figure:

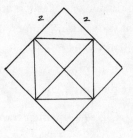

Plato chose to infer the soul's immortality from having a mortal slave recollect an immortal truth; we infer that to this end he himself recalled what he knew from Babylonian lore. Socrates draws a square of side 2 and has the boy deduce that its area is 4. When he asks him for the side of a square with area 8, the boy proposes that the side should be double that of the first square (making a square of four 2×2 squares). Disabused of this by Socrates, he is then led to see that the square on the diagonals of these four squares has half their area, $\frac{16}{2} = 8$.[19]

Notice that Plato chooses his question so that the boy's erroneous middle step—constructing a square of area 16—will lead to the insight, even though the answer won't be the more fundamental $\sqrt{2}$, but $2\sqrt{2}$.

The Old Babylonian tablet also seems to set a leading but much less interesting question, for the inscribed "1 24 51 10" is in a small, neat hand, as is a "30" along one side, meaning $30/60$, or $\frac{1}{2}$. In a sprawling, student hand, beneath the constant for $\sqrt{2}$, is written "42 25 35"—i.e., $\frac{42}{60} + \frac{25}{3600} + \frac{35}{216000} = 0.707106481$: exactly half of the constant which the teacher will have found in one of those pedagogical helps they had even then—a 'table of coefficients'. Can't we hear the teacher asking: if this is the diagonal when the side is 1, what will it be when the side isn't 1 but $\frac{1}{2}$? A relatively easy question, but in a setting which might have been meant to prepare for deeper issues.*

---

* It will not have escaped you that YBC 7289 is a special case of Guido's proof (making all of his four boxes congruent)—and you marvel again at how narrow is the gap between what should and what did generalize, and

What matters for us is that this diagram—scaled up by Plato, down by whoever the teacher scribe may have been*—*shows* by its simple symmetries that the area of a square built in this way on what Socrates calls "the line from corner to corner" must be 2: half of the enclosing square's 4. In neither case are triangles, or the Pythagorean Theorem, invoked (although Plato would have known the Theorem very well).

Our reconstruction, we admit, is by the enjoyably fallacious *post hoc ergo propter hoc*, with antecedents taken as causes (since Plato did this, a Babylonian must have done the same, before him). We warned you that our telling would be postmodern. But is the argument as far-fetched logically as it is in time? If we have similar footprints, eons apart though they are, can we not plausibly argue that similar creatures made them?

$$\triangle \quad \triangle \quad \triangle$$

Let's probe how the Old Babylonians came up with that astonishingly accurate approximation for $\sqrt{2}$, to see what this might tell us about the antiquity of the Pythagorean Insight.

When a courtier leaned over to Bach while Frederick the Great was conducting his orchestra and whispered, "What rhythm!", Bach whispered back: "You mean, what rhythms!" These inveterately calculating Babylonians will disappoint us, as you'll soon see, by having devised many ways of evaluating square roots, and producing widely varying answers—that surprisingly accurate one among them. We know that accuracy and application weren't paramount concerns of theirs,[20] but did a variety of answers to a single question mean that for them reality (at least as far as measurement went) had no definite rhythm? Or are we seeing a jumble of different approaches from different times, places, and schools? We'll look at all of these in the context of a right triangle with short side a, long side b, and diagonal c—whether or not the Babylonians failed to

---

between the truths of mathematics, which are a priori, and their discovery, which is synthetic.

* How neat it would be were this Ishkur-Mansum, or Ku-Ningal, servant of the moon-god Sin, the first mathematicians whose names we know.

separate such a figure from a completing rectangle; and we'll test each against that value on YBC 7289 for $\sqrt{2}$.

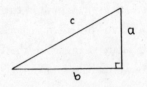

A fascinating late Old Babylonian fragment (recognized in the 1980s to be made up of two yet smaller pieces in different museums)[21] contains two ways* of trying to find each of a, b, and c, given the other two, with a always equal (in our number system) to 10, b to 40, and c taken as 41 ¼. Perhaps it tells us more than it meant to.

The first method takes $c = b + \dfrac{a^2}{2b}$. This isn't a particularly good approximation. For $a = 10$ and $b = 40$ it does indeed give us $c = 41.25$, which isn't far from the correct 41.231 . . . , and for our right triangle with $a = 1$ and $b = 1$, $\sqrt{2}$ comes out as 1.5. But this is the least of the problems with this method. It is meant to yield a and b as well. Given b and c it does indeed produce $a = 10$. But for b there is a real difficulty: as the problem on the tablet reads (translating to our numbers):

> The breadth is 10, the diagonal is 41 ¼. What is the height?
> You: No.

---

* A third method was reconstructed by Christopher Walker (who discovered that the two tablets belonged together) and Eleanor Robson, from what Robson describes (in "Three Old Babylonian Methods", 56–57, 63–64) as "tiny fragments only". This third method is supposedly the Pythagorean Theorem itself. Since the theorem can't give a way of approximating, and since, as Robson comments, "it is impossible [from the condition of the fragments] to determine which approximation to the diagonal was used here", this reconstruction doesn't enter into our thinking—especially since it makes up out of whole clay 48 percent of the text (in translation) of one problem, 72 percent of the second, and 100 percent of the third. Unlike Miss Froy's name, no breathing on these fragments will, it seems, make their message reappear.

This machine can't be run backward—at least by them: they would have to have applied the famous quadratic formula (discovered at least two, indeed more like three, millennia later) to get

$$b = \frac{c \pm \sqrt{c^2 - 2a^2}}{2}$$

This lack of reversibility—and hence of generality—carries a message for us, which we'll soon read.

The second method on this tablet is $c = b + 2a^2b$: a very poor approximation indeed, and one with which, again, they don't even try solving for b—although we know that $b = \dfrac{c}{1 + 2a^2}$. For their $c = 41.25$, this method produces the absurd $.2052\ldots$, and in a right triangle with legs 1, the hypotenuse would be 3 rather than $\sqrt{2}$. Was this formula a scribal error—but if so, for what? Robson generously points out that "its accuracy is dependent on the size of the triangle, as well as its shape . . . the error is less than 10% for triangles with hypotenuse between 0.53 and 0.72."[22] This message goes into the same envelope with the last.

Since a subtractive companion for the first method above, $c = b - \dfrac{a^2}{2b}$, shows up long after, in Alexandria, some scholars would conjure it up in Babylon.[23] It gives the useless estimate ½ for $\sqrt{2}$. All three of these methods, however, are one-shot affairs, so were not only makeshifts in the absence of insight, but couldn't possibly approach what we know are irrational square roots, like $\sqrt{2}$. For this, iteration from a first good guess is needed. The Babylonians must have had this approach to get the value for $\sqrt{2}$ on the Yale tablet (just when you're ready to give up on them, they startle you with their ingenuity). We find three fascinating candidates if we read back, once again, from what was for them the distant future.

The most famous of these ways of closing in on any square root is Heron's Approximation. Make a rough guess, x_1, for the square root you seek, and then find your second approximation, x_2, by calculating

$$x_2 = \frac{x_1 + \dfrac{2}{x_1}}{2}$$

How close, and how quickly, will this get us to $\sqrt{2} = 1.414212962$, if we start with the reasonable guess $x_1 = 1$, and in general take

$$x_{i+1} = \frac{x_i + \dfrac{2}{x_i}}{2} \text{ ?}$$

Although for the Old Babylonians this would have involved division into the "irregular" number 17 (a number whose divisors aren't 2, 3, or 5), they could have managed it, and after five tedious iterations would have gotten 1.414213562: agreement to the first five places. Each further iteration deals with progressively larger fractions, but brings the over-estimate steadily down. You wonder, though, just how much in their style is Heron's Approximation. Why took ye from us our loved Old Babylonian Box?

A second adventure into post hocery brings us to Theon of Smyrna, who lived a generation after Heron, from about A.D. 70 to 136. You'll see his clever idea most clearly if you begin with our isosceles right triangle with both legs $a_1 = 1$ (so we'll call both a_1), and suppose for starters that its hypotenuse c_1 is also 1.

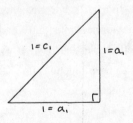

Now extend each leg by $c_1 = 1$, giving us legs $a_2 = 2$, and $c_2 = 2a_1 + c_1 = 3$.

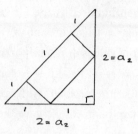

So

$$\frac{c_2}{a_2} = \frac{2a_1 + c_1}{c_1 + a_1}$$

And in general,

$$\frac{c_{i+1}}{a_{i+1}} = \frac{2a_i + c_i}{c_i + a_i}$$

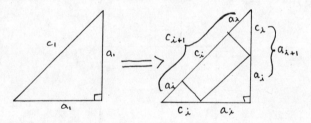

It may not surprise you that successive approximations take on the same values as Heron's, but half as fast, so that Theon's estimates stroll in a more leisurely way toward the Yale tablet's value for $\sqrt{2}$. Can we picture the Babylonian scribes having had the time for these peregrinations? Yes: their society was bureaucratic, but lacked efficiency experts. Will they have had the patience, though, to carry out ever more tedious computations? Surely for each, weariness will at some point have made them hear their mothers calling—without wondering whether or when (this year, next year, sometime, never) the process would conclude. They were very unlikely to have thought of reality itself as shaky: that would have to wait thirty-seven hundred years, for enough angst to kick in. But "approximation" may not have meant to them what it does to us. They may have thought of values differently arrived at the way we think of variant spellings: there is no ideal that 'color' or 'colour' better real- izes. In any event, unlike the problem with triangles whose sides are, say, 10 and 40, the square built on $\sqrt{2}$ gave them the advantage of al- ways being able to test their latest value simply by squaring it. Once again, however, Theon's cleverness doesn't seem theirs (the diagram of it we gave was produced by a nineteenth-century German mathe- matician).[24]

We've saved for last the magnificent idea of Ptolemy (who lived about half a generation after Theon, and not all that far away, in Roman Egypt)—not only because it gets the exact value on YBC 7289 in *three* iterations, but because its method comfortably derives from the Babylo- nian Box![25]

Say you wanted to find $\sqrt{4500}$. Make a square box with that side—

$$\sqrt{4500}$$

4500

whose area is therefore 4500—and fit into its upper left-hand corner the largest square with an integer side that you can. Since the largest integer with a square less than 4500 is 67, that will be the side of our square, and its area will be $67^2 = 4489$; we draw the square, and see that the area of the remaining gnomon is $4500 - 4489 = 11$.

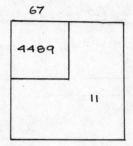

Now fit into this gnomon a smaller gnomon, hugging the square we just drew in, whose width (let's call it $^x/_{60}$) will have x be the largest possible integer that will still keep this new gnomon inside the big, initial square. We obviously haven't drawn this to scale: we've made the 4489 square *much* too small, in order to leave room for the gnomons coming up.

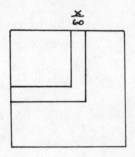

Digression: are you bothered that the width is $^x/_{60}$, not just x? Where did that denominator of 60 come from? We're getting successively better approximations to $\sqrt{4500}$, having found the first approximation $\sqrt{4500} \sim 67$. We would now want to find the next decimal place $\sqrt{4500} \sim 67.x$, where the integer 'x' lies for us in the tenths place, hence stands for the numerator in $^x/_{10}$. But since the Babylonians used 60 instead of 10 as the basis for their place-value system, their successive denominators will be powers of 60.*

---

* We do the same thing in effect when finding the decimal form of, say, $^3/_7$: 7 doesn't go into 3, so we make it 7 into 3.0 and divide as if it were 7 into 30.

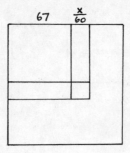

As you see, the area of this gnomon will be that of two rectangles, each $x/60$ wide and 67 long, plus the area of the corner square, $\left(\dfrac{x}{60}\right)^2$. In other words, we want $2\left(67 \cdot \dfrac{x}{60}\right) + \left(\dfrac{x}{60}\right)^2 < 11$.

Good—except that if we are Old Babylonians, we don't know how to solve that quadratic for x! We therefore do something characteristically clever and simply *ignore* the term $\left(\dfrac{x}{60}\right)^2$, since it contributes so very little to the new gnomon's area.* If we therefore solve for x in $2\left(67 \cdot \dfrac{x}{60}\right) < 11$, we find $x < 4.9$, hence we take $x = 4$, so that our better approximation for $\sqrt{4500}$ will be $67 + \dfrac{4}{60}$.

We repeat this process to find y for the next term: $\dfrac{y}{60^2}$, i.e., $\dfrac{y}{3600}$ (see, if you like, the upcoming figure). This will be conceptually easy (being the same process as before) but calculationally hard: just what the Old Babylonians relished.

For what's left of our original square is a slender gnomon whose area is $4500 - (4489 + $ the area of the 'x' gnomon). With $x = 4$, its area (includ-

---

* A similar benign neglect comes up for us in calculus, when we show with such boxes what the derivative of a product of functions is. Here again we get a little square, where two rectangles overlap, and cavalierly ignore its area because it contributes negligibly little to the whole—how long the legs of ancient invention.

ing the small corner square) is $2\left(67 \cdot \dfrac{4}{60}\right) + \left(\dfrac{4}{60}\right)^2 = 11 - \dfrac{536}{60} - \dfrac{16}{3600} = \dfrac{7424}{3600} \sim 2.062$.

How *very* OB: a brain-boggling battle with fractions, to win a mind-bending war of ideas.

So now we want an integer y such that $2\left[\dfrac{y}{3600}\left(67 + \dfrac{4}{60}\right)\right] < 2.062$

—having again ignored the negligible term $\left(\dfrac{y}{3600}\right)^2$. With suitable effort we find $y=55$, so that our improved estimate is $\sqrt{4500} \sim 67 + \dfrac{4}{60} + \dfrac{55}{3600}$.

Go on like this, if you really want to, for the next term, $\dfrac{20}{60^3}$. This will give $\sqrt{4500}$ accurate to five (of our) decimal places.

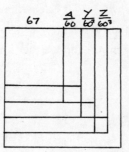

Stepping back, you see not only how brilliant but how very Old Babylonian Ptolemy's method is (like Heron and Theon, he was reputed to have been in touch with old knowledge). Only one last snag has to be removed if we are to apply it—as someone, so long ago, must have—to come up with that little clay tablet's approximation to $\sqrt{2}$.

For what if, at the very outset, the first square you fitted into the box whose side you wished to calculate was so small as to leave a gnomon with an area greater than or equal to its own? You would be stymied. This is the problem that confronts us in approximating $\sqrt{2}$, since the largest square with an integer side that will fit into a square with area 2 has area 1, leaving a gnomon of area 1:

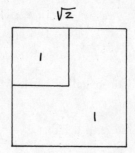

Babylonian cleverness to the rescue again: instead of dealing in sixtieths, multiply your whole calculation by 60, thus avoiding the problem with too big an initial box. As long as you remember to divide your answer by 60, you'll be safe.

We'll therefore begin not with a square of side $\sqrt{2}$ but $\sqrt{2 \cdot 3600} = \sqrt{7200}$, and area therefore 7200. We now proceed exactly as above, but fortunately with easier calculations.

We fit into this square the largest whole number square we can. Since $84^2 = 7056$, we put this in and see that a gnomon of 144 is left.

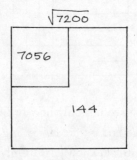

We now have to remind ourselves that this is really $\dfrac{84}{60} = 1 + \dfrac{24}{60}$, or, in the way we transcribe Babylonian, 1;24. The Babylonians just wrote 1 24. Look familiar? This is how the approximation in that palm-sized Yale tablet began—and how it continued follows from just what we did above. If you'd like the details, see the appendix to this chapter.

△ △ △

We have wandered among a variety of Old Babylonian tools for extracting square roots—some of them blunt, some showing the signs of having lain so long in the earth, some still with a bright edge on them. Our faith in imagining, if not reasoning, backward has been affirmed by what we've found in three Middle Eastern mathematicians of the first century A.D.

What has this told us about ancient Babylonian culture, and hence about what forms the precursors of the Pythagorean Insight took among them? It is time to read the messages left for us by those two ungainly methods. Robson remarks of the second (which yielded such absurd results) that it may have remained in use *because of the ease with which it could be calculated.*[26] This sounds like looking for your lost key under the lamppost—a damning judgment on the vast, ramshackle society, taking their lack of concern with utility and application not as idiosyncratic but idiotic. Yet might it not show just what their working *on examples*, rather than *from exemplars*, did: a suiting of each How to the lovely impertinence of What. In this context general truths, like the Pythagorean Insight, were of as little value as they are to cooks, who shop by smell and choose by touch and use the pot with the broken handle because they know from long experience exactly how it will behave.

We said at the beginning that theirs seemed a culture of algorithm—but in fact it was founded on recipes, which are always modified by locality and detail. Bakers need to know which flour they're using, the humidity in the room, and its height above sea level, not to mention the temperature of their own hands. The Old Babylonian scribes had to know which particular numbers were in question and exactly what shape they lay in; they neither thought about nor cared for the generality that most tellingly characterizes mathematical structure (resting on the vital differences between *each* and *every*, *next* and *all*).

The Pythagorean Theorem is a signpost on our way toward the ever more general, and the Babylonian treatment of it—so sensitive to examples—shows vividly how they drifted away from this direction. Yes, they generalized means (such as their brilliant box), but not ends. That sensitivity to initial conditions, you'd have guessed, could only move thought forward. Yet as today's physicists and the invention of fractals have stunningly shown us, its product is—chaos.

APPENDIX

We left our small slab with its approximation to $\sqrt{2}$ at 1;24. To go on, we want to fill up this gnomon as best we can with a smaller gnomon, w wide, whose parts will be two rectangles of area $w \cdot 84$ each, plus a small square in the corner, with area $\left(\dfrac{w}{60}\right)^2$. Its total area will therefore be

$$\frac{2 \cdot w \cdot 84}{60} + \left(\frac{w}{60}\right)^2$$

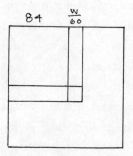

As before, we drop the small square $\left(\dfrac{w}{60}\right)^2$, which we can't deal with and don't need anyway. Since this gnomon must have an area less than 144, we want the largest w such that

$$\frac{2 \cdot w \cdot 84}{60} < 144$$

$$2 \cdot w \cdot 84 < 144 \cdot 60$$

$$w < 51\frac{3}{7}$$

and we are compelled to choose $w = 51$. Our approximation now is $\dfrac{84}{60} + \dfrac{51}{60^2}$, or (in the way we transcribe Babylonian) 1; 24, 51.

If you feel you have aged a lifetime in the last few pages, it is in the good cause of time-machining yourself back to Babylon. Here's the last stage of the journey. Our remaining area is now

$$144 - \frac{2 \cdot 51 \cdot 84}{60} - \left(\frac{51}{60}\right)^2$$

which simplifies to .4775. We now want an integral width y for our next gnomon so that its area will be less than or equal to this remainder—i.e., so that $2 \cdot y\left(84 + \frac{51}{60}\right) + \left(\frac{y}{60^2}\right)^2 < .4775$, and so y must be less than 10.129 . . .

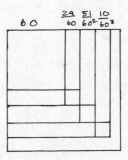

Hence y is 10, and our third approximation is

$$1 + \frac{24}{60} + \frac{51}{60^2} + \frac{10}{60^3}, \text{ or } 1; 24, 51, 10,$$

exactly what you find incised on YBC 7289.[27]

CHAPTER THREE

Through the Veil

Something happened—but not what you'd expect. The theorem we've been vainly chasing through the Middle East emerged, armed with a proof, from the Ionian Sea, more than midway through the sixth century B.C. It wasn't just about right triangles whose sides were in the proportions of 3-4-5 or 5-12-13: it was about infinitely many right triangles, all at once.

Were the course of human affairs rational, this proof would have been the one you saw Huxley put in the hands of his Guido, since it follows so readily from the Old Babylonian play with gnomons in a box (simply divide the gnomon's rectangles into triangles, move them around as they moved around other pieces—and there you are). You could then have argued that someone, in the last days or aftermath of their empire, seized on the pleasure they took in generalizing techniques and carried it to its natural end of generalized insights—and it spread abroad. But in fact the proof that surfaced wasn't anything like Guido's.

Let's take literally that "rational" in our feigned reconstruction and insert Guido's proof in the thousand-year void of evidence between the Old Babylonians, with whose math we now feel ourselves on familiar terms, and this new culture rising in Magna Graecia.[1] Heath gives cogent grounds for doubting this was the proof that the Pythagoreans had.[2] If you label those two termini, a millennium apart, as a and b, the shadowy existence of this proof is an x, making the equality of ratios $\frac{a}{x} = \frac{x}{b}$. Guido's proof is, as it were, a "mean proportional" between two knowns, a way of letting the distant ends approach one another.

You might think that doing this was a kind of smoothing out, something between story and history.* But our supposition has an aim different from making a real series of events approximate to the ideal. By keeping the Guido touchstone at hand as a wholly different proof takes shape, we will see more vividly how radical was the shift from the Old Babylonian viewpoint. It gives a context for asking not only why Guido's proof wouldn't have harmonized with the new cast of thought, but why there had been any proof at all, rather than simply asserting or exemplifying $a^2 + b^2 = c^2$.

Who brought this concern with mathematics and proof in general, this striking shard of it in particular, to a distant outpost of civilization?

Pythagoras, born on the island of Samos, fetched up as a young man in Croton (modern Crotona in Calabria) around 530 B.C.[3] He lived on the cusp between myth and legend, from which the slide into either is sudden and steep. Had he previously been wandering in the wild lands north of Thrace, where the Hyperborean Apollo was worshipped? Or had he been the Hyperborean Apollo himself? Had he been also, or instead, in India, or in Egypt? Perhaps in all, and at the same time, since he was known to be capable of appearing in several places at once.[4] The common theme in the many tales about him, however, is his theory of the soul's migration through the veil of mortality. For the body to be simultaneously in different parts of space has magic to it, but for the soul to travel through time, taking on successively this persona and that, rings of divinity, since it appeals to our deeper Finnegan desires.

Yet what has this to do with the theorem that bears his name, or mathematics generally? The story, like Robert Frost's woods, is lovely, dark, and deep. It takes some telling, because while there is more evidence than there was for the Babylonians, it is obscured by the secrecy so dear to the Pythagoreans and the tangle of rival claims made by their followers and detractors. The solvents and tracer dyes of modern textual analysis have done much to let us look through the verbiage, and with the tweezers and microscopes of scholarship, the rough pieces of evidence

---

* So a Russian linguist, Alexander Lipson, playfully concocted "hypothetical roots" of irregular Russian verbs, to make learning the language easier.

begin to fit into a picture. A striking feature is this: Pythagoras felt compelled to *prove* his theory of metempsychosis.

Why wouldn't a shaman's authority have sufficed? Perhaps because what the eye can't see the heart may not grieve for, but the mind will continue to worry about. And perhaps the brotherhood that began to form around Pythagoras—first at Croton, then 150 miles away, along the coast, in Tarentum (modern Taranto)—already had those skeptical traits for which the Greeks were famous. Not too skeptical, though, since the proofs that convinced them consisted in his recalling publicly some of his past incarnations. While in Argos, for example, Pythagoras saw hanging in a sanctuary the shield of the Trojan hero Euphorbus (which Menelaus had taken when he killed him, half a millennium before), and recognized it as his own.[5] How he knew that he had also once been a peacock, and a son of Hermes, and the commissary Pyrandus (who had been stoned to death), and a Delian fisherman, and the beautiful prostitute Alco, has not come down to us; but since he had also once been Aethalides, the Argonauts' herald whose soul could forget nothing, his power of recollection seems not unreasonable.[6]

At this point human rather than superhuman traits advance the story. Had Pythagoras been nothing more than a shaman with a very good memory, what would there have been to do but worship him—perhaps as a reincarnation of Apollo? But if the transmigrations of his soul meant that ours too survived death, then a cult of hope could form around him; and if the soul was embodied in lives as various as his had been, perhaps one of these hopes could be that the right practices would make the next incarnation better than the present one. The right practices: worship turns into ritual, satisfying our need to do something active for our salvation.

Apollo was the god of catharsis—purification—so these rituals took the form of ever more strict observances to purge your life of the impure. Always put on the right shoe first; do not dip your hand into holy water, nor travel by the main road; food that falls from the table belongs to the heroes: don't eat it. Never sacrifice a white cock; don't stir the fire with a knife; don't break bread; don't look in a mirror by firelight; always pour libations over the cup's handle; don't turn around when crossing a border;

don't eat beans, nor sit on a bushel measure.[7] Hundreds of these "acusmata"—things heard—filled the days of the growing Pythagorean brotherhood with anxiety. You just couldn't be careful enough: demons were lurking everywhere to trip you up. On the other hand, ascetic demands (initiates had to keep silent for perhaps as long as five years)[8] and repetitive practices might induce not only formal feelings but a trance-like state, in which your previous lives were recalled. They would certainly bind the community together in proud exclusivity, as these observances became fixed rules of life.

Although the holiness of the Pythagoreans made them the political power of Magna Graecia, defeating the opulent rival city of Sybaris,[9] at the same time a split developed within the community. The pure never feel themselves sufficiently pure. A clue to something more profound seemed to lie in numerical acusmata. Each number had its own significance: 1 was being, 2 was female, 3 male, 4 justice, 5 marriage, 10 the perfect number, since it was both the sum of 1, 2, 3, and 4, and the harmony in which the Sirens sang—and so on. Some of them therefore found that thinking about number and shape extracted the mind from the material world: imperfection was bypassed, exceptions gave way to rules, this particular number and that to even and odd, examples to exemplars. Our architectural instinct surfaced in them and became central, revealing that the soul which took on various material shapes was ultimately *form*. Studying form was thus the true purification.

Since intense study in general, and in particular of structures rather than things, lessens the sense of self and so seems to free soul from body or let it wander from one to another, this group of *mathematikoi*, as they were called—we'll call them Knowers—looked down on the *acusmatikoi*— the Hearers—as people who obeyed without understanding. The Hearers may have practiced at night recalling the day's events, in hopes of strengthening their powers enough to remember previous lives, but the Knowers studied eternal geometry, number theory, music.*

---

* A companion distinction may have been at work here as well, between the spoken and the written word. The former plays a key role in the wonders of

This set the paradigm for what has become our ingrained distinction between opinion and knowledge (later elevated to knowledge versus wisdom), whose two strands were eventually synthesized in Plato's theory of recollection.

Even more profoundly, contemplating pure form encouraged among the Knowers a sense of distinct souls belonging to a One, and developed the metaphor of a reality behind appearances. When a philosophy emphasizing Being emerged, they would be ready for it. The evolution of their thought from extracted selves to abstract unity—from the soul's reincarnation to its immortality—took the Pythagoreans past the shaman Pythagoras. This accounts for the startling paradox delivered by modern scholarship: Pythagoras had nothing to do with the theorem that bears his name! The only theorem he can rightly claim kinship with is Stigler's: no scientific discovery is named after its original discoverer.*

But haven't we ancient authority for Pythagoras sacrificing a hundred oxen to celebrate his proof? Lewis Carroll (writing as his alter ego, the Reverend Dodgson) first questioned this: "Sacrificing a hecatomb of oxen—a method of doing honor to Science that has always seemed to me *slightly* exaggerated and uncalled-for. One can imagine oneself, even in these degenerate days, marking the epoch of some brilliant scientific discovery by inviting a convivial friend or two, to join one in a beefsteak and a bottle of wine. But a *hecatomb* of oxen! It would produce a quite inconvenient supply of beef."[10]

The problem with Ancient Authority is that—unlike mathematics—it aims to establish truth by deducing important results from worthy

---

bardic recitation, incantatory magic, and secret doctrine; the latter in the greater complexity and abstractness of argument that it bears and promotes. These Pythagorean events took place at a time when traditions were shifting from the former to the latter: the spoken word was indeed in the beginning, but the written in the end.

* This law, proposed by Stephen Stigler in his 1980 "Stigler's Law of Eponymy" (in *Science and Social Structure: A Festschrift for Robert K. Merton*, ed. T. F. Gieryn [New York: New York Academy of Sciences, 1980], 147–57), was named by Stigler after himself because, he says, it isn't originally his.

names rather than from worthy premises. Those numerical acusmata may well have been sayings of Pythagoras, and the view of numbers as archetypal symbols, read for their cosmic significance, might indeed be his. But what had he to do with the games that numbers play in and for themselves?

Pythagoras emerges from current studies as a charismatic cult leader, who could, for example, miraculously tell his followers what had happened among them during his three-year absence (because he had been hiding in a basement, where his mother passed notes down to him about each day's doings).[11] The river Cosas did not rise as he crossed it and hail him by name.[12] He didn't bite a deadly serpent to death in Tuscany, nor stroke a white eagle in Croton. He never foretold the advent in Caulonia of a white she-bear.[13] These studies have even taken from him the golden thigh he displayed to the audience at the Olympian Games, and he hasn't a leg left to stand on.*

△ △ △

Demolishing the reputation of Pythagoras is the new orthodoxy, and in its zeal even seeks to devalue the work of the Knowers. What has been attributed to them likely came instead from the Greek world at large, this argument runs. That "likely", however, reflects more fashion than fact, since we usually can't tell at this remove who conjectured and who proved what. Something we might call Travelers' Internet surely played a larger role than we tend to give it credit for: that spread of chatter via seafarers and adventurers that brought Babylonian doings through the veil to the Western world, and revealed what Thales in Miletus or Hippocrates from Chios was thinking.

When we *can* follow the course of rumor, we see the Knowers

---

* There are four lives of Pythagoras (as befits a man of many incarnations), each more fantastic than its predecessor (see www.completepythagoras.net for Guthrie's translation of these four lives, and all fragments). You will find a paragraph-brief life in Burkert, 165, valuable as much for what it tells us about the character of modern revisionism as for what it says about its subject.

(where, in one another's presence, invention was on the boil) at the focus of this more diaphanous community, as in their struggling over the problem of how, given a cubical altar, to construct another with twice its volume.

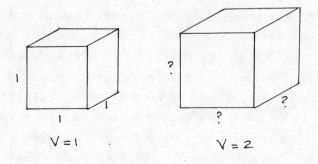

This was asked by Apollo through his oracle at Delos; partially solved by Hippocrates years later and a hundred sea miles away; then completed by Archytas in Tarentum another seven hundred miles and some fifty years distant from him.* What will soon very much bear on our story is the nature of Hippocrates's approach. A prevalent mathematical concern was how to make a square with twice the area of a given square (as we would say, how, given a^2, to find the side of a square with area $2a^2$). They saw that this was equivalent (again, in our terms) to finding a mean proportional between a and $2a$—an x, that is, such that

$$\frac{a}{x} = \frac{x}{2a},$$

---

* That the problem was then differently solved by Archytas's pupil Eudoxus, who came from Cnidos, a thousand miles away in the Anatolian Peninsula; and after, differently still, by Eudoxus's pupil Menaechmus, from another town in Asia Minor five hundred miles farther from Cnidos, shows how broad was the reach of information, and how cosmopolitan the Pythagorean community.

for then $x^2 = 2a^2$, and $x = \sqrt{2}\,a$. Working by analogy, Hippocrates suggested that for doubling a cube, you would need to insert *two* mean proportionals, x and y: If

$$\frac{a}{x} = \frac{x}{y} = \frac{y}{2a},$$

then $ay = x^2$ and $xy = 2a^2$, so $\dfrac{a}{x} = \dfrac{x^2}{2a^2}$, giving us $x^3 = 2a^3$. We would therefore have $x = \sqrt[3]{2}\,a$. We would have it, that is, if we knew how to construct such a length—and it was this that waited fifty years for Archytas and his successors. Keep this story in mind. Keep in mind too that "keeping in mind", rather than before his eyes, was just what Hippocrates did with his proto-algebra. For these manipulations with mean proportionals are invisible; the eye only sees symbols, the mind what they stand for.

Walter Burkert, the leading light in the field of Pythagorean scholarship, makes a sharper attack on the Pythagorean brotherhood. Distinguishing between people engrossed in numerology (shorthand for the metaphorical acusmata) and people engrossed in mathematics, he argues that none of the former would have been able or even have wanted to practice the latter.

How right Burkert would be, were the society of mathematicians made in the image of our last century's logical positivists, those steely scientists so pure of purpose and honed by logic as to be all but computers draped in flesh. You need only look at the variety of weekend practices that actual mathematicians engage in, however, to be assured that their nature, like ours, is a twisted wood that binds reason together with passion. Their weekday lives, as well, are shot through with longings, insights, superstitions, far-fetched conclusions drawn from wild premises, irrational hope and rational despair.

Well, but you think: *as* mathematicians they are surely immune to mystical excesses. We offer you the Name Worshippers. Those who founded the early twentieth-century Moscow School of Mathematics, Dmitrii Fedorovich Egorov and Nikolai Nikolaevich Luzin, belonged—even during the materialistic Soviet Union—to a secret group that believed in the profound power of chanting the Jesus Prayer (made up only of the names Jesus, God). Franny, in J. D. Salinger's novel *Franny and Zooey*, explains it well: "If you keep saying that prayer over and over again—you only have to just do it with your lips at first—then eventually what happens, the prayer becomes self-active. Something happens after a while. I don't know what, but something happens, and the words get synchronized with the person's heartbeats, and then you're actually praying without ceasing. Which has a really tremendous, mystical effect on your whole outlook. I mean that's the whole point of it, more or less. I mean you do it to purify your whole outlook and get an absolutely new conception of what everything's about."

For these Russian mathematicians, this new conception inextricably mixed free will, set theory, redemption, and discontinuity, and out of this mixture came powerful results in the theory of functions, descriptive set theory (where naming became a creative force), probability, and many more fields of mathematics.* The great figures we know from the Moscow School, such as Alexandrov and Kolmogorov, were Luzin's students—and name worshipping continues among the school's descendants to this day.

It really isn't surprising that a mind brimming with conjecture should have the analogies astir in it take their symbols as standing for external events as well as for one another, conjoining the mathematical and the

---

* You may rightly wonder how discontinuity, free will, set theory, and redemption could possibly combine. Pavel Florenskii, the mathematician and mystic behind the Name Worshippers, attributed the ethical decline of the nineteenth century to its infatuation with calculus, built as it was on a faith in continuity, which in turn, he saw, established the world as deterministic (differential equations showed what *must* follow from initial conditions). But Cantorian set theory had demoted continuous sets to but one of many varieties. With discontinuity restored to the worldview, determinism fell, free will was now possible, therefore religious autonomy and, beyond causality, redemption. For the fuller story, see Loren Graham and Jean-Michel Kantor's *Naming Infinity* (Cambridge: Belknap Press of Harvard University Press, 2009).

metaphysical (the gnomon doesn't just extend, it embraces the square it belongs with): that the childish and the childlike should seamlessly mingle, whether in Tarentum, Moscow, or Harvard. But Burkert's claim that the Knowers would not have wanted to prove anything follows from what he takes to be obvious, that proofs diminish the high romance of cloudy symbols. He writes: "That which fills the naïve mind with amazement is seen [by alert mathematical analysis] as tautologous and therefore self-evident."[14] This peculiar view belongs to what we'll call Russell's Syndrome, after Bertrand Russell, who wrote in his autobiography: "I set out with a more or less religious belief in a Platonic eternal world, in which mathematics shone with a beauty like that of the last Cantos of the *Paradiso*. I came to the conclusion that the eternal world is trivial, and that mathematics is only the art of saying the same thing in different ways."[15]

This appalling misunderstanding of mathematics comes from having failed to notice that saying the same thing in different ways, like the slight difference in the angles of our eyes, gives stereoptical depth. The profundity of mathematics follows from its taking two different insights as equal. Why else would a great contemporary mathematician be overcome with appreciative joy every time he revisits a proof of the Pythagorean Theorem?[16]

△ △ △

So how did the Knowers in Tarentum prove the theorem that honors their founder's name? Their lives were pervaded by music and their imagination by its cosmic power: not only did the stars circle harmoniously (we fail to hear their music only because it has been in our ears from birth),[17] but all things are "locked together by harmony" (as the late Pythagorean, Philolaus, put it) "if they are to be held together in an ordered universe."[18] Their mathematics, accordingly, was steeped in proportions, since they had discovered that the chief notes of the scale came from the simplest whole number ratios. Pluck a string at its midpoint and it sounds the octave above (they expressed this half as the ratio 2:1). Stopping it in the ratio 3:2 produces the fifth, and 4:3 the fourth. There were those numbers again, 1, 2, 3, and 4, that made up 10,

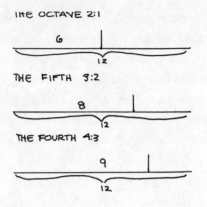

their talisman *tetraktys*.

Ratios were therefore on their minds when they looked at what had filtered through from Babylon, the 3-4-5 right triangles and their scaled-up versions. Since scaling preserves ratios (as the Babylonians, or anyone dealing with plans, would have known),* it would have been natural to draw these similar triangles *without any accompanying numbers,* that one might stand for all. They let their thoughts loose on these generic figures: a momentous instance of less being more. Turning them

---

* It was certainly known to Thales, some fifty years before these events, who determined the height of the Egyptian pyramids by comparing the length of their shadows to that of a measured stick he had set up. The significance of this rang through the isles of Greece: the world afar behaves like the world at hand.

this way and that, as their thoughts turned, a scatter of different sizes suddenly harmonized:

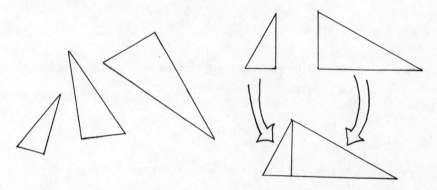

That vertical line, the join of two smaller right triangles within the larger, similar third, takes on a life of its own, since it looks just like the proportional stopping of a plucked string—and also like a mean proportional.* But the use of *two* mean proportionals was in the air—whether because Hippocrates's work on doubling the cube had preceded these thoughts, or followed like them from some prior inspiration. Watch how, instead of using these two means in series, they let them play together.

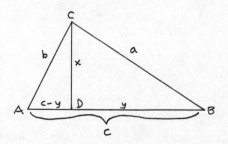

The two triangles, ΔADC and ΔCDB, are *similar*: the second, that is, is the first simply scaled up; their corresponding angles are the same, and

---

* It seems clear (Burkert, 440–42) that the Pythagoreans were familiar with the arithmetic and harmonic as well as the geometric means. Only the first two arise in divisions of a one-dimensional string ("the cutting of the canon"), the third bringing us, as you see, into the second and third dimensions.

their sides are in proportion. The symbol for "similar" is, understandably, like a weakened "=": ~. Since $\triangle ADC \sim \triangle CDB$,

$$\frac{c - y}{x} = \frac{x}{y}$$

and we say that x is the mean proportional between $c - y$ and y. We have, from this equation, that

$$x^2 = cy - y^2. \tag{1}$$

But also (because $\triangle ACB \sim \triangle CDB$) a is the mean proportional between c and y:

$$\frac{c}{a} = \frac{a}{y}$$

so
$$a^2 = cy. \tag{2}$$

Substituting (2) in (1), $x^2 = a^2 - y^2$, that is,

$$a^2 = x^2 + y^2,$$

in right triangle CDB—and hence, similarly, in all right triangles.

The flash of an insight rarely has a single source (what is the light from one stick rubbing?), and Guido's proof may have been involved, if it had indeed been invented and circulated by this time: for it is only a few hops of the imagination from moving four congruent right triangles around in a square box to fitting two similar right triangles into a triangular one (lengthening that square box to a rectangle and then cutting it along its diagonal would have been the first such hop).

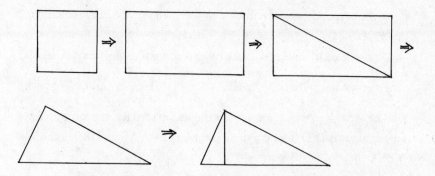

In any case, the Knowers wouldn't have put their proof algebraically as we did, but would likely have thought of it in terms of areas, drawing on the Old Babylonian manipulations that showed through the veils of distance and time. This is how it might have gone, pairing a different mean proportional involving the length b, rather than x, with a.

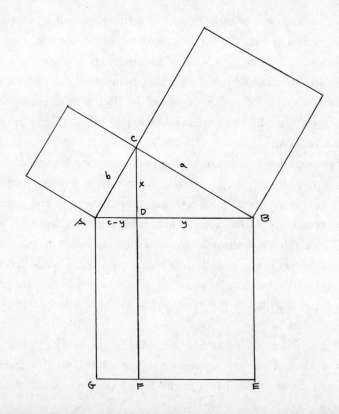

Drop the perpendicular CF from C to the far side, GE, of the square on side AB.

Since ∠CDB and ∠ACB are congruent right angles, and ∠CBD is congruent to itself, ΔCBD ~ ΔABC, so $\dfrac{c}{a} = \dfrac{a}{y}$. Hence, multiplying both sides by ay, $a^2 = cy$, which means that the square on a has the same area as the rectangle BEFD (for note that c is the length of AB and therefore also of BE).

Similarly, ΔCAD ~ ΔBAC, so $\dfrac{c}{b} = \dfrac{b}{c-y}$, and, multiplying both sides by $b(c-y)$, $b^2 = c(c-y)$.

That is, the square on b has the same area as the rectangle AGFD.

Since the area of these two rectangles adds up to the area of the square on c, the theorem is proven.

△ △ △

A proof convinces us by showing that something apparently surprising or unlikely lies in a matrix of relations we recognize and truths we acknowledge. So Pythagoras's proof of his numerous reincarnations tied them to objects we could see (the shield of Euphorbus), names we knew (the Argonauts' herald) and figures from daily life (a peacock, a fisherman, a prostitute). By this criterion, Guido's proof and that by proportions are on a par.

But from this criterion of harmonious connection a finer one soon emerges: not just the magic of seeing the astonishing object mirrored among its contemporaries (like Pythagoras appearing in several places at once), but seeing it derived from the foundation of things: deduction imitating descent. It is immeasurably more meaningful that Pythagoras reincarnated Apollo than Alco. By this higher standard, Guido's proof is dazzling prestidigitation: now you don't see it, now you do. But the proof by proportions strikes to the inner eye, which sees not appearance but structure.

Music and mathematics had combined to justify faith in the harmony of things. Only one step remained to attach this proof to the cosmic frame. Since for the Pythagoreans the universe was made of whole numbers and their ratios, and the ratios are manifest in this proof, just call up now their constituent numbers.*

Ingenuity had taken these numbers away in order to gain the flexibility of generic figures, so that the numbers, having yielded this vital insight, could now be restored.

Just as the octave and all the musical intervals are built up from a unit tone, so here, the a, b, c, x, and y in a right triangle must have a common measure. If a and b are both 1, c must be a number which harmonizes with 1: a whole number—or, if not, a ratio of whole numbers, m/n, so that when we scale up the triangle's sides by n, we get n, n, and m. But, if a and b are 1, by this very theorem $c = \sqrt{2}$. We therefore need to find the ratio—or, some would say, the fraction—m/n, such that $\sqrt{2} = \dfrac{m}{n}$. This need ushers in the last act of our drama—tragic or comic, according to your lights. Travelers' Internet would very likely have brought to Tarentum those brilliant approaches the Old Babylonians took to evaluating $\sqrt{2}$, like the one we saw reincarnated in Ptolemy. There just three iterations gave $\sqrt{2}$ as

$$\frac{1414212962}{1000000000}$$

But if we recall how we got that value, it takes little sophistication to see that it might not be the final one. In fact, the amount of space left in the L-shaped room after putting in the first three gnomons makes it certain that more work remains. Worse: since successive gnomons clearly take up less and less area, the merest shadow of a doubt about *ever* being done flickers across the mind. If this mind is already disposed toward

---

* Was there a harbinger to this Pythagorean summoning of number by form in the Old Babylonian play between shapes—their geometrical gnomons— and the numbers that came from filling them up with pebbles? We certainly recognize, in this discarding of a means once the end it leads to is in sight, a process that plays out again and again in mathematics (think of the epsilon-delta approaches to a limit in calculus)—along with its opposite, the means replacing the importance of the original end.

skepticism (as we suggested before that Greek minds were), doubt can take on a far more vital form and even raise the unthinkable question: might the problem not lie in the method but in $\sqrt{2}$ itself: might it not be a ratio of whole numbers at all?

Perhaps this question wasn't so thoroughly unthinkable. A rival musical theory existed at the time, which saw harmonic intervals not as ratios but as lengths of a plucked string: a viewpoint more congenial to awkward—even to *continuous*—divisions.[19] Of course, it would have taken a mind open to ideas from a camp toward which the Pythagorean brotherhood was actively hostile. Something begins to materialize behind the veil—or someone, more shadowy still than Pythagoras: the hero or villain of the piece, Hippasus.

We hardly even know his dates—some time around 450 B.C., perhaps. That way the inner disruption would have neatly coincided*[20] with the brotherhood's sudden and surprising loss of political power in southern Italy—when rivals burned down their house at Croton and the only survivors, two young Pythagoreans, fled. We do know what emerged from Hippasus, or from someone taken to be him: we know the proof that $\sqrt{2}$ couldn't be a ratio of whole numbers. Again it took rubbing together two sticks, one of which came over the water: the Babylonian inheritance of the deep dichotomy between even and odd numbers. The interest the Pythagoreans took in these was primarily for their mathematical worth (such as square numbers being the sum of successive odds), but with value added on by other meanings: odd was male, light, good, square, and right, even was female, darkness, bad, oblong, and left.

---

* But was even the internal crisis wholly due to Hippasus? The stories about him are cloudy. Was he drowned at sea by the gods for his impiety in having discovered (or just disclosed?) the irrationality of $\sqrt{2}$? Was it instead the irrationality of the Golden Section within the sacred pentagon that he revealed? Or did he more thoroughly subvert the teachings of Pythagoras or, Oedipus-like, betray and displace him? Did he lead the Knowers against the Hearers, only to be abandoned even by them and called a plagiarist? Or was it he who came to understand through theory and experiment the mathematics of musical concord? The utter discord among ancient accounts of Hippasus is well presented in Burkert, 455–65.

The other stick traveled overland from Elea (modern Velia), a town a hundred miles west of Tarentum on the Tyrrhenian Sea. It was here that Parmenides devised the philosophy which, as we remarked earlier, the Pythagoreans would be ready for when it came: a philosophy devoted to the opposites of being and non-being.[21] His message was as simple and profound as a tautology: being is, non-being isn't. This recognition gave rise to the wholly new sort of proof by contradiction, based—like evens and odds—on opposition: if you assume something *is*, and following where your assumption leads lands you in something which *can't be*, then your initial assumption must also not have been. An upbringing exposed to the numerous contradictions among the acusmata, as well as those in the life and teachings of their founder, would have helped prepare the mind for this way of using the power of contradiction.

Here is how Greek skepticism, possibly Oedipal rivalry, the even-odd distinction, and proof by contradiction combined into one of the great monuments to our power for distinguishing with the mind what the eye could never see, and for answering by a single act of reason what no amount of experiment could ever establish. We will decant its spirit into modern bottles.

THEOREM: $\sqrt{2}$ isn't a ratio of whole numbers.

PROOF:

Assume $\sqrt{2}$ is such a ratio, i.e., that $\sqrt{2} = \dfrac{m}{n}$.

Had m and n a common factor, cancel it out before proceeding, so that m/n is now in lowest terms (m and n, therefore, can't both be even).

Squaring both sides, $2 = \dfrac{m^2}{n^2}$.

Multiply both sides by n^2: $2n^2 = m^2$.

Hence m^2 is even, and therefore (since even times even is even but odd times odd is odd), so is m. We therefore write $m = 2a$, for some number a.

Substituting $2a$ for m, $2n^2 = (2a)^2$, or

$2n^2 = 4a^2$.

Dividing both sides by 2, $n^2 = 2a^2$, so that n^2 is even and therefore so is n.

Hence m and n are both even, contradicting our second step.

We were therefore wrong in assuming that $\sqrt{2}$ was a ratio of two whole numbers.

A proof as elegant, headlong, and devastating as anything in Sophocles. The incommensurable had torn the veil of numerical harmony, and with it the harmony of the soul, society, and music itself.[22]

A deeper understanding of what had been achieved by this proof would have to wait until almost the present: a wait exemplified by a young Guido of our acquaintance, who once drew on the blackboard an isosceles right triangle with legs of length 1. After having worked his way through Hippasus's proof, however, he returned to his diagram and solemnly erased its hypotenuse, since it didn't exist. He had yet to learn the deepest lesson of Parmenides: rather than *not being* rational, $\sqrt{2}$ *is* irrational. Being Is.*

---

* But *what* it is becomes less, not more, comprehensible. Not only are the creatures making it up more various and complex than we had thought, but our ways of coming to know them grow ever more subtle. So, in this story, we have passed from diagrams with specific lengths to generic diagrams— and now, with proof by contradiction, to no diagrams at all, where we are on the brink of algebra, with saying replacing showing. Our architectural instinct carries us toward transparent structure—toward structure per se, because (as the Pythagoreans' contemporary Heraclitus wrote, and you read at the outset), a hidden connection is stronger than an apparent one. This is the ultimate meaning of the veil through which we see.

CHAPTER FOUR

Rebuilding the Cosmos

Euclid alone has looked on beauty bare.
—EDNA ST. VINCENT MILLAY

There is no excellent beauty that hath not some strangeness in the proportion.
—FRANCIS BACON

The eruption of the irrational shattered the Pythagorean cosmos and littered the ancient Greek landscape with its fragments. Whether or not Hippasus had hoped to take over the brotherhood, his $\sqrt{2}$ banished the Pythagorean vision of harmony from the Theban landscape of thought. It took little further effort to see that $\sqrt{2}$ wasn't the sole exception to the rule of rationality: any rational multiple of it would be irrational too (for if k were such a multiple, and $k\sqrt{2}$ could be expressed as a/b, then $\sqrt{2} = a/kb$, making $\sqrt{2}$ rational—which we know to our sorrow it can't be). And soon Theodorus of Cyrene showed that $\sqrt{3}, \sqrt{5}, \sqrt{6}, \sqrt{7}, \sqrt{8}, \sqrt{10}, \sqrt{11}, \sqrt{12}, \sqrt{13}, \sqrt{14}, \sqrt{15}$, and $\sqrt{17}$ were irrational too, and then his student Theaetetus proved that irrationality was the rule: all square roots, save those of perfect squares, were irrational.

This was the mathematical chaos inherited by Greece at its acme: the serpent under the skin of that classical order you see in the thought of Plato, the statesmanship of Pericles, the works of the Athenian dramatists, architects, and sculptors. In each, however, discordant elements were reconciled: "harmonized opposites," as Heraclitus put it, "as of the bow and the lyre." After an epoch of exploration, with its triumphs, shocks, and confusions, mathematics needed time to make a stable foundation on which the shaken structure of number and shape could once again settle. Sense had to be made of the mathematical enterprise itself, and this in turn required a deeper understanding of proportion: one that would include irrational quantities too—not only to save the brilliant proof of the Pythagorean Theorem which you saw in the last chapter (and which depended on a theory of proportions dealing only with ratios of whole numbers), but to show how we may think through time about eternal things.

Eudoxus, who studied with Archytas and then briefly with Plato, came up with a way of handling proportion that answered these needs. It was as sophisticated as any of our own abstract devisings. Rather than asking what a proportion *was*, and whether there had to be a common measure among its terms, he laid down rules for how it must *behave*, whatever it was. To assert that $\frac{a}{b} = \frac{c}{d}$, he said, meant that for any whole numbers m and n, if $ma > nb$, then $mc > nd$. If $ma = nb$, then $mc = nd$. But if $ma < nb$, then $mc < nd$. These rules of conduct were all we knew, and all we needed to know, about proportion.

If you find this unnerving, so would the contemporaries of Eudoxus. When Euclid, the Great Consolidator, put together his towering *Elements* toward the end of the turbulent fourth century, it took the whole of his Book V to work out the properties and consequences of this definition. His aim was to build mathematics up from self-evident truths, yet the Eudoxian view of proportion was anything but self-evident. Since Euclid needed the Pythagorean Theorem quite early on,[1] in order to prove a number of fairly basic propositions, how could he do so, if the revised proof by proportions had to wait until the theory behind it stabilized, close to halfway through his work?

Perhaps it is only when you see human affairs in the middle distance that they make conventional sense. Farther back they become too hazy to decipher; closer to, they are utterly strange. In his steadily growing monument to mind, with its wealth of common notions, definitions, postulates, lemmas, porisms, and hundreds of propositions, Euclid proves the

Pythagorean Theorem *twice*. The second time, in Book VI, is indeed by proportions—but at the end of Book I he gives a proof renowned for its ingenuity and notorious for its difficulty, which avoids proportions entirely.

Sir Thomas Heath long since pointed out that Euclid's approach was nevertheless inspired by the Pythagorean. You might have expected instead that the proof would be like Guido's, especially since Books I and II of the *Elements* are filled with the Babylonian Box. It isn't—but we could see it as drawn from the spirit of that proof as well. Why hadn't Euclid used such a simple, elegant, and intuitive approach, which needs no more than a "Behold!" to convince you? Because his aim was to found mathematics forever—and in response to the Eleatic imperative, described in the last chapter, its pictures were to be of perfection, its objects grasped as unchanging. Triangles can't be shunted around in a box, configured now this way, now that: these figures belonged to Being, and were therefore at rest. And just here, in our view, Euclid's imagination triumphed: he let the figures remain still, but moved mind among them!

How? By comparing their areas.* If you want to show that the area of a given square, for example, is equal to that of a certain rectangle, find a pair of congruent triangles (whose areas would therefore be equal) and prove that one of them has half the area of the square, the other half that of the rectangle. This transitivity of area, through the triangle's middle term, was in the spirit of a mean proportional,

---

* This idea may well have stemmed from the Pythagoreans too. The Greek historian of mathematics Eudemus—not much younger than Euclid— speaking through the later writer Proclus, tells us that constructing a figure with a certain area on a given straight line, called "the application of area", was "one of the discoveries of the Muse of the Pythagoreans". Might Euclid have taken even more of this proof from the Pythagoreans? For it is associated with applications of areas in Plutarch's *Symposium*: "This [application of areas] is unquestionably more subtle and more scientific than the theorem which demonstrated that the square on the hypotenuse is equal to the squares on the sides about the right angle." Quoted in Heath's *Euclid*, I.343–44.

while its differently disposed triangles you could think of as derived from Guido-like manipulations; it was now, however, a legitimate act of thought, moving over the waters of the world.

Having understood this to be Euclid's strategy, his proof of proposition I.47 becomes clear. Yet there is a veil of clarity that falls between it and the reader. In reaction to Pythagorean mysticism, the Greeks saw that a proof's authenticity must follow not from rhetoric or its maker's reputation, but from no more than mechanical inevitability (the *ananke* of their drama). They set what is still our style of passive imperative, virtually impersonal, mathematical narration, with all traces purged of the imaginative work and insights that went into a proof's invention (the passive imperative becomes a perfect passive imperative when it comes to constructions: "Let this *have been* constructed," thus solving once again the Eleatic problem that mathematical objects must be at rest; so they are, the constructing having occurred in some prior world, before the proving began). These *formal* proofs are held to clockwork standards, the gears seen to turn and mesh of themselves. Such are the proofs that seem so intimidating on the pages of most texts, where imagination is assumed to be at its 4 A.M. low ebb. Here is Euclid's, rearranged organically.

I.47:

In ΔABC, right-angled at C, the sum of the areas of the squares on the two legs (AHKC and CLMB) equals the area of the square on the hypotenuse (ABEG).

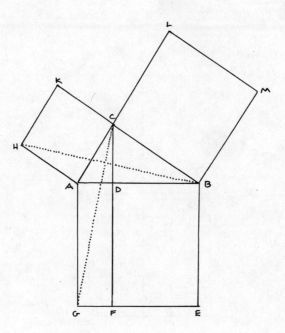

PROOF:

Part I.

1. Drop the perpendicular from C to AB, meeting it at D, and extend it to meet GE at F (there's the legacy from the Pythagorean proof by proportions).

2. Consider ΔAHB. The altitude to its base HA from B is equal in length to AC (think of HA extended, making a parallel to KCB. The altitude from B is then the perpendicular, equal in length to AC, between these two parallels).

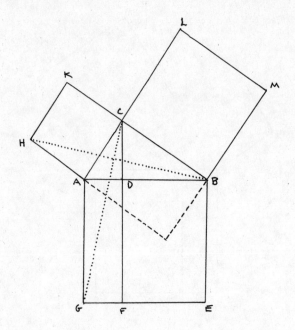

3. That gives the area of △AHB, half the base times the height to that base, as $\left(\dfrac{1}{2}\right)$HA · AC.

4. Hence △AHB has half the area of square AHKC.

5. Consider △ACG. Using "≅" to mean what its two parts imply—not only similar but equal—in length, in angle measure, or in shape, AC ≅ AH, AG ≅ AB, and ∠CAG ≅ ∠HAB (each is a right angle plus ∠CAB), so △ACG ≅ △AHB, by side-angle-side, hence their areas are equal.

6. But reasoning as in step 2, the area of △ACG is $\left(\dfrac{1}{2}\right)$AG · AD.

7. Hence (by transitivity of areas), the area of AHKC equals that of rectangle AGFD.

Part II.

Proceed exactly as in Part I, considering now △ABM and △CEB:

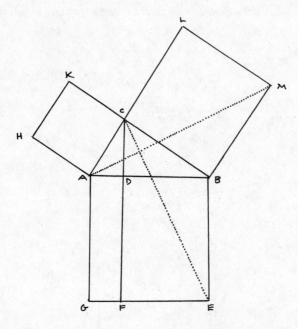

1. As in the steps 2–4 of Part I, △ABM has half the area of square CLMB.
2. As in step 5, △ABM ≅ △CEB.
3. As in step 6, the area of △CEB is $\left(\frac{1}{2}\right)$ EB · BD.
4. As in step 7, the area of CLMB therefore equals that of rectangle BEFD.

Since the rectangles AGFD and BEFD together make the area of ABEG, adding the results of Parts I and II gives us the sum of the areas of the squares on the two legs (AHKC and CLMB) equaling the area of the square on the hypotenuse (ABEG), as desired.

Of course the diagram with which Euclid confronts his reader presents both parts at once, and with HB, CG, AM, and CE (which, being "auxiliary lines", *we* dotted in) shown as solidly as the rest; so that the first thing the reader sees is this cat's-cradle, with its pairs of obtuse triangles, reminiscent of Guido's four right triangles, caught as if in a double exposure:

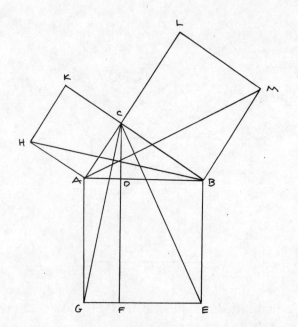

No wonder Schopenhauer, mixing his metaphors in the throes of a warp-spasm, called this "a mouse-trap proof" and "a proof walking on stilts, in fact, a mean, underhanded proof" (the German *hinterlistiger* sounds even more villainous).[2]

Worse: don't those sharp angles suggest a dangerous and barely caged creature? If you've ever wondered why we speak of the "horns of a dilemma", your answer is here. In his 1570 edition of Euclid, Sir Henry Billingsley (who later became a Haberdasher, then Sheriff, then Lord Mayor of London) wrote that I.47 "hath been commonly called of barbarous writers of latter time Dulcarnon", from the Persian *du-*, two, and *karn-*, horned[3]: the dilemma that impales whoever would tackle this proof.

"I am, til God me bettre minde sende,

At dulcarnon, right at my wittes ende."

That's Criseyde, speaking two centuries before in Chaucer's *Troilus and Criseyde*. Yet those who conquered this proof came to venerate it. Proclus, writing in the fifth century A.D., says: "If we listen to those who wish to recount ancient history, we may find some of them referring this theorem to Pythagoras and saying that he sacrificed an ox in honor of its discovery. But for my part, while I admire those who first observed the truth of this theorem, I marvel more at the writer of the *Elements* . . . because he made it fast by a most lucid demonstration."[4]

"Made it fast": Proclus uses the same image of tethering that Plato puts in Socrates's mouth when kidding Meno about the supposed statues in Athens made by Daedalus, which he pretends would run away were they not tied down (by way of explaining to Meno that true opinions only become knowledge when tethered by recollection). This tells us something important about the Greek view of diagrams, constructions, and proofs, in terms of their attitude toward what *is* versus what we *know*: that distinction between ontology and epistemology which has ever since split the Western mind as surely as the Great Longitudinal Fissure divides the brain's two hemispheres.

For the image of tethering "the truth of this theorem" (i.e., making it more than just a true opinion) by a demonstration shows that these mathematicians were as aware as we are that it takes constructive work to grasp what lies in being, rather than becoming. Euclid's constructions (he wrote three books, now lost, about them)[5] are mean proportionals between being and thinking: scaffoldings to be removed once the finished work is revealed.* Kant, more than a millennium later, rethought this as the *synthetic* and the *a priori*, the made and the found, recognizing them as two sides of the same coin. At the end of a demonstration Euclid presents us with everything, constituent and auxiliary lines alike, on

---

* Is this the origin of what is by now a common mathematical trope? You will find it everywhere after, as in the epsilon-delta shims that square up calculus.

a par. The end of his work is the beginning of ours, who must, like readers of Plato, rebuild his artifices and then, with him, remove them.*

We can now understand why those awkward obtuse triangles came to give this theorem its other, gentler, medieval nickname: the Bride's Chair. Do you see the maiden luxuriating there as an Eastern bearer carries her palanquin on his back to the wedding? Heath discovered her once again in a Great War issue of the magazine *La Vie Parisienne,* where Euclid's diagram was the framework for a French *poilu* shouldering his bride and all his household belongings.[6]

This jolly misogynist version went back to an inn sign painted by Hogarth, *The Man Loaded with Mischief.* Heath caught a glimpse too of such a sign somewhere in the fen-land near his own Cambridge.[7] Those days are perhaps farther from ours than are Euclid's.

△ △ △

It isn't quite true that Euclid proves the Pythagorean Theorem twice: when he returns to it in Book VI he actually proves a strikingly broader statement.† For in developing Eudoxus's theory of proportions, he shows that not just the areas of squares but of *any* similar polygons on a right triangle's three sides will be in the relation $a^2 + b^2 = c^2$, where a, b, and c are the side-lengths of these fitted polygons.

As you'll soon see, Euclid's proof will glide as smoothly as a swan—but the vigorous paddling beneath the surface will both deepen the story of intermediate constructions and shift Euclid from a name

---

* The intermediate status of Euclid's constructions—these *beings* that occupy a middle ground between Being and Becoming, allowing mind to do the moving—is shown in several ways. The miniature Greek drama of each of his propositions begins with a general statement, which is then repeated in a ghostly lettered diagram that, being unscaled, is generic, and made following instructions for ideal rather than real instruments; and always in the chorus, an echo of proofs by contradiction that can't have diagrams. The drama ends with the return of the general statement, risen from its ghost.

† We attribute these proofs throughout to Euclid, when in fact he may have no more than anthologized the work of many others—who live, nameless, through him.

55e Année, N° 5   Le Numéro : 60 centimes   Samedi 3 Février 1917

LA VIE PARISIENNE

Un débrouillard ou   le sac a Malice

on a staid classic to an almost companionable fellow being, by showing us the cast of his thought if not the cut of his jib. For he makes what we'll call *Euclid's Converter* by breaking up similar polygons, as we would, into networks of similar triangles, and then showing (see the appendix to this chapter) that the areas of similar triangles are to each other as the squares of their corresponding sides—hence so must be the areas of their sums. But it is just here that we have the advantage of him, being the proud possessors of algebra, the ultimate in generic reasoning.

In algebraic style we would have no diagrams whatsoever but call our similar triangles Δ_1 and Δ_2, and denote their areas by $|\Delta_1|$ and $|\Delta_2|$.

Since we know that a triangle's area is half the base times the height, $|\Delta_1| = \frac{1}{2}\, b_1 h_1$, and $|\Delta_2| = \frac{1}{2}\, b_2 h_2$.

If $\Delta_1 \sim \Delta_2$, with a constant of proportionality k (so that $b_2 = k b_1$ and $h_2 = k h_1$), then

$$\frac{|\Delta_1|}{|\Delta_2|} = \frac{\frac{1}{2} b_1 h_1}{\frac{1}{2} b_2 h_2} = \frac{\frac{1}{2} b_1 h_1}{\frac{1}{2} k b_1 k h_1} = \frac{1}{k^2}.$$

But since $b_2 = k b_1$, $k = \dfrac{b_2}{b_1}$, hence $k^2 = \dfrac{b_2^2}{b_1^2}$, and so $\dfrac{1}{k^2} = \dfrac{b_1^2}{b_2^2}$. Therefore $\dfrac{|\Delta_1|}{|\Delta_2|} = \dfrac{b_1^2}{b_2^2}$. The areas of similar triangles are proportional to the squares of their corresponding sides: a little postprandial walk through the forest of symbols.

Since Euclid had general figures, not abstract ones, he could reach this result only by some clever geometric manipulations—and we know from I.47 which way his mind inclined: toward making an intermediate triangle that would allow transitivity of areas!

Given $\triangle ABC \sim \triangle DEF$, with $\triangle ABC$ the larger, what Euclid did was to choose a point G on BC so that $\dfrac{BC}{EF} = \dfrac{EF}{BG}$. Contortions worthy

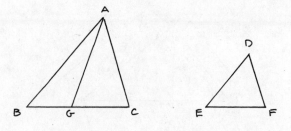

of a Houdini struggling out of undersea chains then led him to conclude that $|\triangle ABG| = |\triangle DEF|$, and later, that $\dfrac{|\triangle ABC|}{|\triangle ABG|} = \dfrac{BC^2}{EF^2}$, so that $\dfrac{|\triangle ABC|}{|\triangle DEF|} = \dfrac{BC^2}{EF^2}$, as desired. The mathematician Peter Swinnerton-Dyer once remarked that there is a "sort of mathematics that no gentleman does in public."[8] You might well think Euclid's machinations are of this sort, so lest you run screaming at the sight, or at that of the apparatus in his mind's gymnasium, we'll hide them in the appendix at the end of this chapter.

Now at last we're ready for Euclid's enhanced Pythagorean Theorem, in modern notation:

THEOREM: In $\triangle ABC$, right-angled at C, if the polygons P_a and P_b on the two legs are each similar and similarly situated to that on the hypotenuse, P_c, then the sum of their areas equals the area of that on the hypotenuse: $|P_a| + |P_b| = |P_c|$.

PROOF:

[representing any such triple of similar polygons by similar parallelograms, and using "$\Rightarrow$" to mean "implies"]:

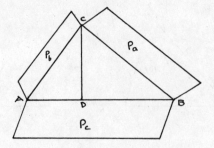

1. Drop a perpendicular from C, meeting AB at D.

2. $\triangle CAD \sim \triangle BCD \sim \triangle BAC$

    (1)   (2)   (3)

    (by corresponding pairs of congruent angles)

3. From triangles (3) and (1), $\dfrac{AB}{CA} = \dfrac{CA}{AD}$,

4. so by (A) in the Appendix, taking $a = AB$, $b = CA$, $c = AD$,

$$\frac{AB}{AD} = \frac{(AB)^2}{(CA)^2} = \frac{|\,\triangle \text{ on } AB\,|}{|\,\triangle \text{ on } CA\,|}.$$

$$\uparrow$$

    (C), in Appendix

5. Likewise, from triangles (3) and (2), $\dfrac{AB}{CB} = \dfrac{CB}{BD}$,

6. So by Appendix (A) and (C), taking $a = AB$, $b = CB$, $c = BD$,

$$\frac{AB}{BD} = \frac{|\,\triangle \text{ on } AB\,|}{|\,\triangle \text{ on } CB\,|}.$$

7. Hence (from step 4), $AD = AB \cdot \dfrac{|\,\triangle \text{ on } CA\,|}{|\,\triangle \text{ on } AB\,|}$,

8. And (from step 6), $BD = AB \cdot \dfrac{|\,\triangle \text{ on } CB\,|}{|\,\triangle \text{ on } AB\,|}$,

9. So, $AD + BD = AB \cdot \dfrac{|\,\triangle \text{ on } CA\,|}{|\,\triangle \text{ on } AB\,|} + AB \cdot \dfrac{|\,\triangle \text{ on } CB\,|}{|\,\triangle \text{ on } AB\,|}$

$$= AB \cdot \left(\frac{|\,\triangle \text{ on } CA\,| + |\,\triangle \text{ on } CB\,|}{|\,\triangle \text{ on } AB\,|} \right).$$

10. But $AD + BD = AB$, so

$$AB = AB \cdot \left(\frac{|\triangle \text{ on } CA| + |\triangle \text{ on } CB|}{|\triangle \text{ on } AB|} \right).$$

11. Dividing both sides by AB, and multiplying by $|\triangle \text{ on } AB|$,

$$|\triangle \text{ on } AB| = (|\triangle \text{ on } CA| + |\triangle \text{ on } CB|) \Rightarrow |P_a| + |P_b| = |P_c|.$$
$$\uparrow$$

Euclid's Converter

This proof legitimizes the first of our theorem's many progeny, waiting in what heaven for their human parents to be born?

APPENDIX

We spoke of Euclid's gymnastics in proving that the areas of similar triangles are to one another as the squares of their corresponding sides. Here are the workout machines, which he needs in the proof of the generalized Pythagorean Theorem.

(A) [This is the modern version of his porism to Proposition 19 of Book V]:

If $\dfrac{a}{b} = \dfrac{b}{c}$, then $\dfrac{a}{c} = \dfrac{a^2}{b^2}$.

PROOF:

$$\frac{a}{b} = \frac{b}{c} \Rightarrow \frac{b}{a} = \frac{c}{b} \Rightarrow \frac{1}{a} = \frac{c}{b^2} \Rightarrow$$
$$\quad\quad\uparrow \quad\quad\quad \uparrow \quad\quad\quad \uparrow$$

invert   times $\dfrac{1}{b}$ multiply on left by a/a

$$\frac{a}{a^2} = \frac{c}{b^2} \Rightarrow \frac{a}{c} = \frac{a^2}{b^2}.$$
$$\uparrow$$

multiply both sides by $\dfrac{a^2}{c}$

Euclid now establishes that the areas of two triangles are to one another as the squares of their corresponding sides. He first proves that

(B) If in $\triangle ABC$ and $\triangle DAE$ we have $\angle BAC \cong \angle DAE$ and $\dfrac{EA}{AB} = \dfrac{AC}{AD}$, then $|\triangle ABC| = |\triangle EAD|$.

PROOF:

(It helps to draw these triangles as here, and then to draw BD).

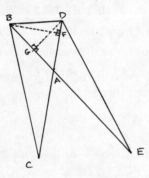

1. Drop $BF \perp CD$.

   Then $\dfrac{|\triangle ABC|}{|\triangle BAD|} = \dfrac{\frac{1}{2} BF \cdot AC}{\frac{1}{2} BF \cdot AD} \Rightarrow \dfrac{|\triangle ABC|}{|\triangle BAD|} = \dfrac{AC}{AD}$.

2. Similarly, drop $DG \perp BE$.

   Then $\dfrac{|\triangle EAD|}{|\triangle BAD|} = \dfrac{\frac{1}{2} DG \cdot EA}{\frac{1}{2} DG \cdot AB} = \dfrac{EA}{AB}$.

3. Hence $\underset{\uparrow}{\dfrac{|\triangle ABC|}{|\triangle BAD|}} = \underset{\uparrow}{\dfrac{AC}{AD}} = \underset{\uparrow}{\dfrac{EA}{AB}} = \dfrac{|\triangle EAD|}{|\triangle BAD|}$.

         step 1 given step 2

4. $\Rightarrow |\triangle ABC| = |\triangle EAD|$.

Now he can prove

(C) VI.19: If $\triangle ABC \sim \triangle DEF$, then $\dfrac{|\triangle ABC|}{|\triangle DEF|} = \dfrac{BC^2}{EF^2}$

PROOF:

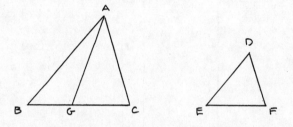

1. Let $\triangle ABC$ be the larger of the two similar triangles. Then

$$\triangle ABC \sim \triangle DEF \Rightarrow \frac{AB}{DE} = \frac{BC}{EF}.$$

2. Choose G on BC so that $\dfrac{BC}{EF} = \dfrac{EF}{BG}$.

3. Then $\dfrac{AB}{DE} = \dfrac{EF}{BG}$.

4. Since $\angle B \cong \angle E$, we have fulfilled the conditions of (B) above, so $|\triangle ABG| = |\triangle DEF|$.

5. Step 2, and (A) above (taking $a = BC$, $b = EF$, $c = BG$), therefore give us $\dfrac{BC}{BG} = \dfrac{BC^2}{EF^2}$.

6. But $\dfrac{BC}{BG} = \dfrac{|\triangle ABC|}{|\triangle ABG|}$ (they have the same altitude from A).

7. So $\dfrac{|\triangle ABC|}{|\triangle ABG|} = \dfrac{BC^2}{EF^2}$.

8. But by step 4, $\dfrac{|\triangle ABC|}{|\triangle DEF|} = \dfrac{BC^2}{EF^2}$, as desired.

Exercise machines? If you look at these and at the rest of his proofs in Book V, they may remind you more of a carpenter's shop before the invention of lathes. There are the beautiful saws and draw knives and files to shape and smooth a chair leg: the work of contemplative hours, with the end-product marked by the craftsmanship of risk. We want the chairs, of course, and many of them, and now—but at the cost of an almost sensuous feeling for form?

CHAPTER FIVE

Touching the Bronze Sky

Some people collect Ketchikan beer coasters, some Sturmey Archer three-speed hubs, others wives or ailments. Jury Whipper collected proofs of the Pythagorean Theorem. He wasn't the first: fifty-nine years before him, in 1821, Johann Joseph Ignatius Hoffmann published more than thirty; in 1778 a Frenchman named Fourrey included thirty-eight among his *Curiosités Géometriques*, and a Herr Graap had translated others from a Russian anthologist. Nor was Jury Whipper the last: Professors Yanney and Calderhead, from the Universities of Wooster in Ohio and Curry in Pennsylvania, gathered together some hundred proofs between 1896 and 1899. A lawyer at the District of Columbia Bar, named Arthur Colburn, published 108 of his own, starting in 1910; perhaps the currents of litigation ran more slowly in the days before air conditioning.

On the shoulders of these giants still stands Elisha Scott Loomis: the boy born in a log cabin in 1852, who rose to be a 32nd degree Freemason and, he tells us, "plowed habit-formation grooves in the plastic brains of over 4000 boys and girls and young men and women." He tells us that of all the honors conferred upon him, he prized the title of Teacher more than any other, "either educational, social, or secret."

What might some of those secret titles have been? And was he as leonine as his portrait shows him, florid moustache and wing collar?

Or should we believe a penciled note in the Harvard Library's copy* of his book, *The Pythagorean Proposition: Its Proofs Analyzed and Classified and Bibliography of Sources for Data of the Four Kinds of Proofs*: "He was somewhat high in manner, but was in reality a good sport. I never met him." Then how did you know, G. W. Evans?

Pythagorean investigations breed mystery. Here's another. How many proofs are there in Loomis's book? Be a Babylonian, you say, and count them. But counting, as even a Babylonian knew, is one of the hardest of human tasks. Loomis claims to have 367 proofs in his second edition, though some are circular, some defective, some no more than

---

* Inside the front cover of this copy is a printed label: "This book was stolen from Harvard College Library. It was later recovered. The thief was sentenced to two years at hard labor. 1932."

variations on or parts of others. Are the algebraic proofs, which he says readily follow from this or that geometric demonstration, to be thought of as different from them? He asks about a possible proof here, "can't calculate the number" of others there, and speaks of "several", "a number of", and "countless" different proofs from those he gives. 9,728 proofs, for example, derive from his figures for Algebraic Proofs Six and Seven, he tells us, and 65,780 more from his Figure Eight. When he writes, as in his note to Geometric Proof 110, of more cases extant, does he mean more than he has given? Are two proofs really different if a square in one has no more than slid sideways from that in another (as in his Geometric Proofs 111 and 112), or if a grid of lines is differently parsed (proofs 119 and 124)?

We conclude that his book contains 367 proofs minus a few, plus several, increased by a number derivable but not in fact derived, to which are added those that are "other" and "different", resulting in many plus a multitude, increased by what he describes as an unlimited horde of the likely, and a quantity indefinitely great that will be discovered by "the ingenious resources and ideas of the mathematical investigator", giving us as an approximate total more than we should, or could, or may, or want to, count. This amplitude is consonant with the generous spirit of brotherhoods, Pythagorean or secret, and is an image of life itself: in the earth below each tree of spreading order the mice of Somewhat gnaw, while chaos in its foliage is made by the insects of Et Cetera.

The Babylonian urge makes us want to number Loomis's proofs; the Egyptian urge, to picture them. His is an anthology of insights had by generals and artists and lawyers and children and principals of normal schools and presidents of free schools; by people from as far away as ancient India and Renaissance Italy and modern England, and others as close to home as Independence, Texas, Hudson, New York, and both Whitwater and Black Hawk, Wisconsin. Among their variety nestle fifty-five proofs by Loomis himself (or is it ninety-four?).

He lets us look a little down the reversed telescope of time to March 28, 1926, which was a particularly good day for him. At ten in the morning on that Saturday he invented a snappy geometric demonstration, and at three in the afternoon another; and then, before he went to sleep, came up with a third, at nine thirty in the evening. And here, twenty-six years before, is Loomis at work on August 1, 1900, during a summer vacation from plowing grooves in the plastic brains of his students at West High School in Cleveland, Ohio. What was a sleepy Wednesday for some meant enough leisure for him to contrive five proofs—and four more for that matter, in the next eight days, after three in the week before (we shall do our best to keep from reading any Pythagorean significance into those numbers). His thoughts were running on figures where the square on a right triangle's short side is folded over onto part of the triangle itself, and onto part of the square built on its hypotenuse. In 1926 it was the square on the hypotenuse that he saw folded back over the triangle. He urges every teacher of geometry to use paper-folding proofs, although the one he cites requires folds we are unable to see.

△ △ △

Might these numerous proofs not reduce in fact to variations on Guido's, or the Pythagorean proof by proportions, with the occasional crossover product such as Euclid's? Does the stark, surprising relation these proofs aim to establish not constrict the flow of invention toward it into a few narrow channels—and at the other end, aren't the sources of our play as challengingly limited as a child's: a stick and a string?

Were you to turn on the news and hear "Now for some baseball scores. Four to three, one to nothing, eight to five . . .", you'd think the world had gone comically mad. This is the madness within whose shadow algebra lives. If some insubstantial part of us, which we variously call instinct and intuition, inclines us this way or that when devising a proof, algebra is the machine in this ghost, churning out, like the mad scientist's Igor, a succession of means without any care for what end they might serve, or whom they should benefit.

Take those 75,508 or so algebraic proofs that Loomis cites. You could tie the triangle up with ever fancier ribbons and then unwrap it,

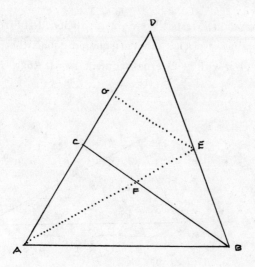

before a familiar construction inside is smothered to death. You could fence it about with palisades and ankle-breakers, keeping ratios in and rationality out.

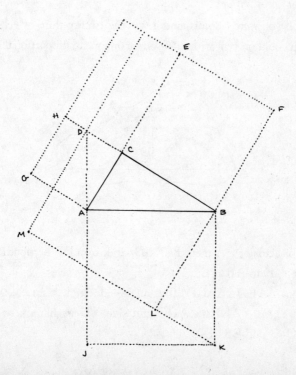

Proofs, more proofs, Master! Were you an aficionado of trig, you could surround it: since a circle's tangent squared equals the product of a secant through it with its external segment, $a^2 = (c+b)(c-b) = c^2 - b^2 \Rightarrow c^2 = a^2 + b^2$.

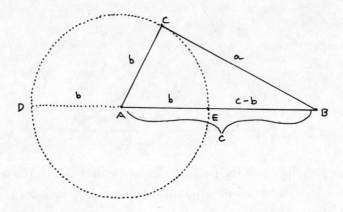

Or, if you were Loomis, and it was the twenty-third of February in 1926, you could prove it without a square or a mean proportional in sight:

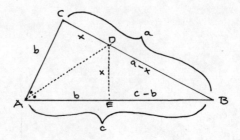

Let the bisector of $\angle A$ meet BC at D, and drop DE perpendicular (or in symbols, $\perp$) to AB at E.

Then $\triangle AED \cong \triangle ACD$ (by angle-side-angle, since $\angle C \cong \angle E$, $\angle A_1 \cong \angle A_2$, so $\angle ADE \cong \angle ADC$ and side AD is shared), so $AE = b$, $DE = DC = x$ and $EB = c - b$.

Since $\triangle ABC \sim \triangle DBE$, $\dfrac{c}{a} = \dfrac{a-x}{c-b} \Rightarrow c^2 - bc = a^2 - ax$, or

$$c^2 = a^2 - ax + bc \tag{1}$$

And

$$\frac{b}{a} = \frac{x}{c-b} \Rightarrow bc - b^2 = ax \tag{2}$$

Substituting (2) in (1), $c^2 = a^2 - bc + b^2 + bc = a^2 + b^2$.

You see in proofs like these a miniature history of the baroque becoming the rococo. Some of Loomis's are tortuous, some teasing; some pert, some monstrous. Yet while most of mental space, like cyberspace, is empty, its convolutions have no room for the convoluted: math worthy of the name is as simple as possible—but not, as Einstein pointed out, simpler.

If algebra pours out such a vacuous plenty, wouldn't trigonometry follow with ampler waves? No, Loomis thunders: "There are no trigonometric proofs, because all the fundamental formulae of trigonometry are themselves based upon the truth of the Pythagorean Theorem; because of this theorem we say $\sin^2 A + \cos^2 A = 1$, etc. Trigonometry *is* because the Pythagorean Theorem *is*." Here, however, is a trigonometric proof deciphered from something sketched on the back flyleaf in Harvard's copy of Loomis (by the same Mr. Evans who told us that Loomis was a good sport):

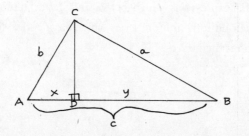

1. $\cos B = \dfrac{a}{c} \Rightarrow a = c \cdot \cos B$, and (in $\triangle CDB$)

   $\cos B = \dfrac{y}{a} \Rightarrow y = a \cdot \cos B$, so, multiplying,

   $ay = ac \cdot \cos^2 B \Rightarrow y = c \cdot \cos^2 B$.

2. Likewise (since $\angle ACD \cong \angle B$), $b = c \cdot \sin B$, $x = b \cdot \sin B$

   $\Rightarrow bx = bc \cdot \sin^2 B \Rightarrow x = c \cdot \sin^2 B$.

3. Adding, and then dividing by c, $\dfrac{x+y}{c} = \sin^2 B + \cos^2 B$.

4. But $x + y = c$, so $1 = \sin^2 B + \cos^2 B$.

5. Since $\sin B = \dfrac{b}{c}$ and $\cos B = \dfrac{a}{c}$, $1 = \left(\dfrac{b}{c}\right)^2 + \left(\dfrac{a}{c}\right)^2$,

   so $a^2 + b^2 = c^2$.

We hear Loomis exclaiming, "One moment, sir! What role has trigonometry played here?" For the proof no more than relabeled (in terms of angle functions) the ratios we knew from the Pythagorean proof in Chapter Three, then withdrew to let those ratios play their familiar game. This seems an instance of tautology yielding nothing new under the central sun of Mind.

A wholly—and profoundly—trigonometric proof blossoms, however, when a great swath of mathematics is rethought as grounded in infinite series (a point of view that—although it puts formalism before intuition—unifies much, solves delicate problems, and opens fresh vistas). For sine and cosine, thought of as functions of their angles, are defined independently of the Pythagorean relation, and their derivatives—the functions that measure their slopes—can likewise be found without it. These lead (via ingenious series devised by an eighteenth-century Englishman, Brook Taylor, in response to a conversation in a coffeehouse) to the wonderful expressions which better and better approximate them, and are at their limit the infinite series which are identical to the sine and cosine functions themselves:

$$\sin x = x - \frac{x^3}{3!} + \frac{x^5}{5!} - \frac{x^7}{7!} + \cdots$$

$$\cos x = 1 - \frac{x^2}{2!} + \frac{x^4}{4!} - \frac{x^6}{6!} + \cdots$$

Squaring these series* and adding the results yields

$$\sin^2 x + \cos^2 x = 1.$$

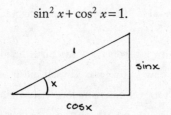

This proof (which lived on the dark side of Loomis's moon) also reminds us of the broader nature of math, by showing that, while it may take a considerable yet finite effort of bookkeeping to keep track of how terms cancel out or vanish, it needs time's infinite expansion to conclude.

Stepping back just an ergo from this flurry of proving, you see that Loomis spoke from an excluded middle, being neither right nor wrong. Proofs of the Pythagorean Theorem via trigonometry are indeed possible—though some are no more than cosmetic changes on proofs closer to the bone, while others coil as infinite series, like stem cells in their very marrow.

△ △ △

If we are to find further significant proofs of the Pythagorean Theorem, we should begin by looking in its proper home, which is geometry: measuring earth. Long before he became famous for his Flashman novels, George MacDonald Fraser had fought in the jungles of Burma, and he describes in his memoirs this strange moment from 1944:

---

* Squaring an infinite series means (we're sorry to say) multiplying all of its terms by each of its terms, and keeping track of what happens. This would ruin the health of mathematical accountants as surely as recording credit and debit wore out Dickensian clerks—were it not for glimmers of pattern and hopes of order along the way: some terms will negate others, and some will dwindle to nothing as the number of terms increases. Here, when the two squared series are ultimately added together, only the 1 is left standing—as you may surmise by taking just the first two terms of each series, squaring them and adding: $\left(1 - \dfrac{x^2}{2!}\right)^2 + \left(x - \dfrac{x^3}{3!}\right)^2 = 1 - \dfrac{2x^2}{2!} + \dfrac{x^4}{(2!)^2} + x^2 - \dfrac{2x^4}{3!} + \dfrac{x^6}{(3!)^2}.$

"You mentioned isosceles triangles. Will it do if I prove Pythagoras for you?"

"Jesus," he said. "The square on the hypotenuse. I'll bet you can't."

I did it with a bayonet, on the earth beside my pit—which may have been how Pythagoras himself did it originally, for all I know. I went wrong once, having forgotten where to drop the perpendicular, but in the end there it was. . . . He followed it so intently that I felt slightly worried; after all, it's hardly normal to be utterly absorbed in triangles and circles when the surrounding night may be stiff with Japanese.[1]

This story not only renews the sense of our ultimate human kinship: it caters to that feeling you get on overcast days, that while the real magic of things casts no shadow, there is an occult meaning foreshadowed by everything. What if Pythagoras really had seen this theorem by scratching it in the earth? Wasn't he after all a traveler through chthonic doorways? So Faust, millennia later, called Mephistopheles up through similar diagrams on the floor, and romantic musings find the irrational caged in all such symbols. And at these times, when significance seems to shine only against a dark background and the gnomon's shadow to tell more than the light that cast it, you find yourself in sympathy once again with the Hearers and the Name Worshippers and the tribe of astrologers and readers of entrails and of Bible Codes, and all references are covert and everything comes to stand for something else.

Let's return to our anthology of proofs, and relish a few of the many flowers that grow in these Pythagorean fields—some as exotic as orchids, some daisy-simple, with a damask rose next to the latest hybrid tea. The Greeks thought that the rare mortal might touch just once, and then but briefly, the bronze sky, through a superlative act of courage, strength, or thought—such as grasping an immortal truth. These proofs of the Pythagorean Theorem, brewing up, who can say how, out of our disorderly daily affairs, may, unrecognized, have marked the acme of their inventors' lives.

Here, for example, is one put together by Thabit ibn Qurra, who was born in 836 in Mesopotamia—two millennia after those Babylonians who did their devising in Chapter Two. Thabit had been a money changer in Harran, but was brought by a chance meeting into Baghdad's House of Wisdom, where he remained to translate ancient Greek texts into Syriac, and to become an expert in medicine, astronomy, astrology, linguistics, magic, philosophy, mechanics, physics, music, geography, botany, natural history, agriculture, meteorology—and mathematics. Think of him when next you look at a sundial: he and his grandson Ibrahim studied the curves that went into its making (such as those you see here in Ptolemy's 'analemma').

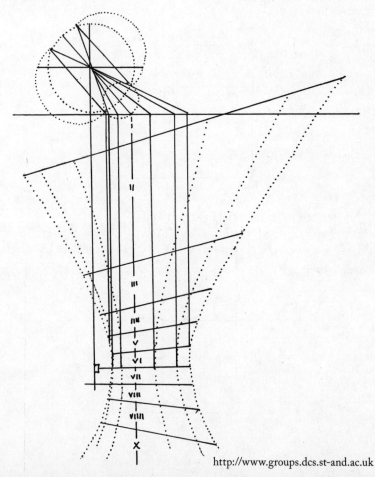

http://www.groups.dcs.st-and.ac.uk

Thabit begins his magical proof by picturing our triangle ABC and its attendant squares, with the small square on side a, and then carrying off those squares and placing them side by side, as squares HBCD and FGDE. Thus ED = b and DC = a.

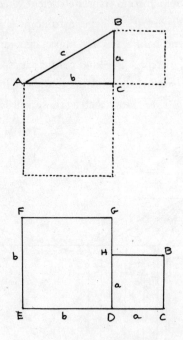

Now he locates a point A on ED so that EA = a. This means that AD = $b - a$, so ADC = $b - a + a = b$. Drawing AB, he has his original triangle back again.

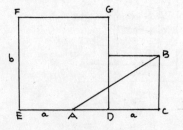

He now draws FA. Let's number the pieces as here:

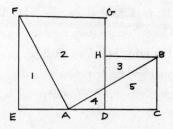

Rotating piece 1 counterclockwise by 90°, it rests on FG as piece 6, with the new vertex K. Of course $6 \cong 1$, so their areas are the same.

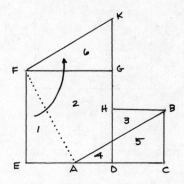

Now rotating $\triangle ACB$ clockwise by 90°, it rests on BH, as part 7. $7 \cong 4+5$, again with the same areas.

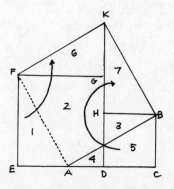

This means that the square on the hypotenuse of $\triangle ACB$ is made up of the pieces $2+3+6+7 \cong 2+3+1+4+5$, which is the sum of the squares on the legs. Done!

Or should we have said done—again? Has Thabit's proof cast a shadow not his, but Guido's, with two of the triangles differently moved and interpreted? For (lettering as below) triangles P and Q remain as they were, outlining the square on the hypotenuse, but S and R have each been rotated 180° counterclockwise, and no longer outline the square but have become part of it!

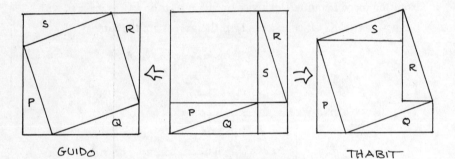

GUIDO   THABIT

Which do you prefer, the simple or the subtle?

△ △ △

And which do you prefer, a proof like Thabit's that involves cutting up, moving, and adding pieces of polygons, or one that uses no more than comparing—with the added fillip of needing the whole Euclidean plane to do so? But if we're not to dissect, then what will we be comparing? Take the two squares with side-lengths *a* and *b* that Thabit began with, set next to each other, and tile the whole plane with them, in every direction, as if this were the Turkish Bathhouse of the Gods. Mathematics is nothing if not extreme.

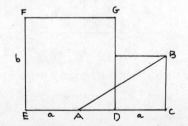

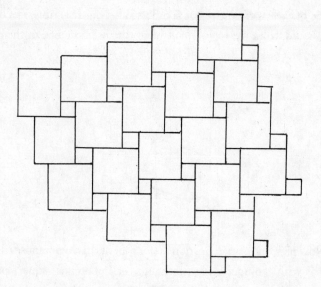

Pause now and pull a transparent layer over this flooring, and on it trace a second pattern—this one just of the hypotenuse squares, but tilted, as in Thabit's diagram:

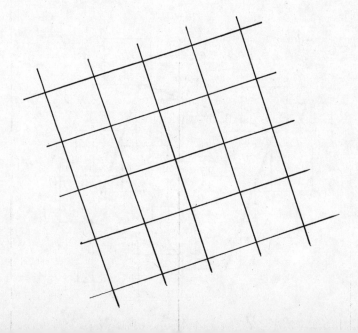

The side of each such square (we'll call its side-length 1) runs, as a hypotenuse should, from the top of an *a* side to the opposite end of the paired *b* side:

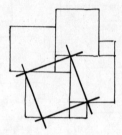

Now pick out a small portion T (for top) of this hypotenuse tiling—here it is with the floor underneath and the hypotenuse square pattern superimposed on it. To see the bigger (but not too big) picture, let's choose to have $n = 3$ hypotenuse squares on a side, so this piece is made of $n^2 = 9$ tilted hypotenuse squares, and it looks as if there were $n^2 = 9$ large and $n^2 = 9$ small square tiles on the floor below it.

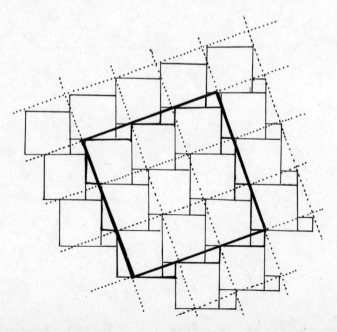

Were this *exactly* so, we would have proven the Pythagorean Theorem: one hypotenuse square would take up as much area as the squares on the two legs. But is it exact? As in our little sample, parts of some of the tiles below stick out of T, and parts of other tiles intrude. Let's not stoop to measure, but call the underneath tiling of n^2 large and small tiles, in our extract, U (yes, for "underneath"), and denote its area—as we did in Chapter Four—by $|U|$. We want to show that $|T|=|U|$, but do this by looking, like gods, from afar—since gods don't sweat the small stuff (even in a Turkish bath).

If the area of our 3×3 patch, T, *fell short* of that of the small and large squares associated with it below in U, it would certainly still fall short in a 4×4 or 5×5 or any $n \times n$ extract. Hence it would fall short if we covered the whole plane in this way, taking the limit as n goes to infinity (in mathematical shorthand, lim as $n \to \infty$)*:

$$\lim_{n \to \infty} |T| \leq |U|$$

(We've written $\leq$ rather than $<$, to hold out hope for the equality we want.)

But if the 3×3 area of T *exceeded* that of U, it would still do so at the limit:

$$\lim_{n \to \infty} |T| \geq |U|$$

That means there must be some constant c, which can do double duty: were T's side-length, n, increased by this c to $n+c$, we and the gods could make a new square that would comfortably enclose U; but also were n shrunk by c to $n-c$, we'd have a square smaller than T which would be enclosed by U. That is:

$$(n-c)^2 \leq |U| \leq (n+c)^2.$$

---

* What n are we talking about? The n which is the side-length of T, so it is covertly there in the letter T, whose area is n^2.

Divide* this inequality through by n^2, and take the limit as n goes to infinity once more—which means letting T become an ever larger square of these 1×1 hypotenuses, while U increases with it, until the entire plane is covered by U and this pulsing T above. So (expanding those squared parentheses and dividing, as we promised, by n^2)

$$\frac{(n-c)^2}{n^2} \leq \frac{|U|}{n^2} \leq \frac{(n+c)^2}{n^2} \Rightarrow \frac{n^2 - 2cn + c^2}{n^2} \leq \frac{|U|}{n^2} \leq \frac{n^2 + 2cn + c^2}{n^2},$$

then, expanding the numerators and carrying out the division by n^2,

$$1 - \frac{2c}{n} + \left(\frac{c}{n}\right)^2 < \frac{|U|}{n^2} \leq 1 + \frac{2c}{n} + \left(\frac{c}{n}\right)^2,$$

which means the limit as n goes to infinity gives us $1 \leq \dfrac{|U|}{n^2} \leq 1$.

Hence $\dfrac{|U|}{n^2}$ is trapped between 1 and 1, so it must *be* 1, and therefore $n^2 = |U|$—that is, T and U have the same area! We have once again (but strangely!) proven the Pythagorean Theorem.[2]

This proof, so different from any we've seen, takes up not only more room but more thought than Thabit's. We've practiced a new sort of trapeze-work here: setting up a less and a greater which turn out to be the same—thanks to some ancient arithmetic and some eighteenth-century taking of limits. You'd be right to feel dizzied at first by this Cirque under a distant Soleil. Yet it would have been a relief to Thabit, since, as you saw, nothing in it moved, save the focal point: out to infinity and back again. Thabit was a firm believer in the Greek view that the objects of mathematics are always at rest, so he must have been troubled by the scrambling around in what he called his "reduction to triangles and rearrangement by juxtaposition".[3]

$\triangle$ $\triangle$ $\triangle$

The style of a proof reflects the character of its maker. Our fictitious Guido's was swift, Thabit's deft, this latest wildly imaginative—but da

---

* This is legitimate, since n is always greater than 0.

Vinci's is cryptic. Construct, he says, the squares on a right triangle's three sides, as lettered here, and then draw EF, making $\triangle CEF \cong \triangle CAB$.

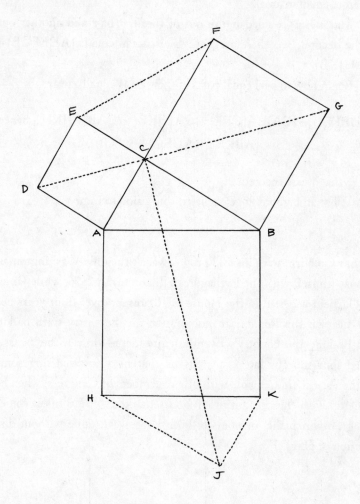

Now construct $\triangle JKH \cong \triangle CAB$ on KH, and last, lines CJ, DC, and CG—an unpromising beginning. At least DCG is a straight line, since its two parts, DC and CG, bisect equal opposite angles.

Leonardo, like a master of misdirection, now asks us to look at four quadrilaterals. After a moment's thought you will see that

DEFG ≅ DABG, and CAHJ ≅ JKBC (in each case a shared side and equal vertex angles). It will take more than a moment to see that DEFG and CAHJ are congruent too, so that all four of them are (and are therefore equal in area).

Making a six-sided figure out of the first pair and another out of the second, we therefore have their areas equal: |ADEFGB| = |CAHJKB|.

Begin now to peel equal parts of their areas equally away:

$$|ADEFGB| - |CAB| - |CEF| = |CAHJKB| - |CAB| - |JKH|, \text{ hence}$$
$$|ABKH| = |ADEC| + |CFGB|,$$

the Pythagorean Theorem.

What ambiguous angel signaled this vision to him?

△ △ △

Arthur Colburn, with his 108 proofs, wasn't the only Washingtonian to busy himself with the Pythagorean Theorem. In 1876, while James A. Garfield was still in the House of Representatives (four years before he won the Republican nomination on the thirty-sixth ballot, and became the country's twentieth president—only to be assassinated four months later), he tells us that he was discussing some mathematical amusements with other members of Congress, when he came up with this elegant proof (and remarked, "We think it something on which the members of both houses can unite without distinction of party.")

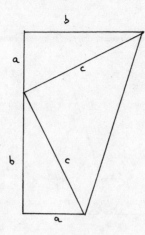

Since a trapezoid's area is its height times half the sum of its bases, the area of this one is

$$(a + b) \cdot \frac{a + b}{2}$$

On the other hand, this trapezoid is made up of two right triangles with legs a and b, and half of the square on side c—so its area can also be thought of as

$$\frac{ab}{2} \cdot 2 + \frac{1}{2}c^2$$

or

$$ab + \frac{c^2}{2}$$

Equating these two expressions for our figure's area gives us $\frac{a^2 + 2ab + b^2}{2} = ab + \frac{c^2}{2}$, or, simplifying:

$$a^2 + b^2 = c^2.$$

Obvious once you see it—like so much math; but (again, like so much math) not at all obvious before. Yet might you not indeed have met it in another existence—or at least in a previous chapter? Isn't it our oldest diagram, sliced slantwise in half?

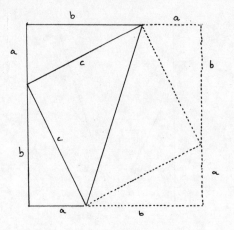

Could some shade of that Greek or Babylonian figure have fallen on Garfield's past, during his 1840s schooldays in Ohio's Orange Township, or at the Western Reserve Eclectic Institute, or while he was at Williams College in Massachusetts? Or is it so deeply lodged in our mathematical genes that an hour spent with convivial colleagues in Washington, or scratching in the Italian dirt, will call it up?

△ △ △

This dizzying variety of proofs must make you want to take stock and classify. The last two examples built outward from the triangle, the Pythagorean inward, and the trigonometric ignored it altogether. We have proofs bounded in a nutshell and others needing infinite space. Would it help to set up kingdoms of algebra and geometry, then phyla of the intuitive and the formal, classes of analytic and synthetic, say, orders of simple and subtle, and families that look to which squares are folded over or folded back? How would we fit in distinctions between proofs without words and proofs without pictures? Should we stick with our anthologist's plan to no more than sample these flowers, or does the Linnaeus in us demand a perfected plenum, which our inner Darwin would then turn sideways and send through time?* Yet classifying rigidifies, and might make us miss the world cavorting beyond our ken.

Would either shelving or browsing explain why there are so many proofs? Is it the fad of an adolescence we pass through or abide in, a beacon of awe that beckons our growing strength? Is it a desire to make the impersonal one's own, or an expression of the astonishment Hobbes felt on first seeing Euclid I.47: "By God! This is impossible!" A reasonable exclamation, since, as we noticed, we haven't the instinct for area that we have for length. This drove him to follow its ancestry back to ever more venerable forebears, as it might us to prior convictions that bear our own likeness on them. While we may sketch these on paper, Hobbes was given to scribbling geometrical diagrams on his thigh (perhaps this is why Pythagoras's was golden). Proofs by proportion, proofs by dissection—perhaps you feel, "Seen one, seen 'em all"—but what if you couldn't see?

THE BLIND GIRL'S PROOF

Loomis credits his fourteenth geometric proof "to Miss E. A. Coolidge, a blind girl." He then gives us no more than a reference to the journal he found it in. What!? Were his curiosity, his imagination, his compassion, not stirred? Did his compulsion to move on to the next, and the next, and the next leave him no time to wonder at the visions of the blind? Not being in such a hurry, we hunted out the *Journal of Education*, volume 28, 1888, through the stacks in Harvard's Gutman Library, where it had—who knows how many years before—been mis-shelved, giving the diligent librarian a dusty two hours before she cornered it down a stack receding to infinity. "That made my day!" she said. And there, on page 17, was Miss Coolidge's proof, across from the "Notes and Queries" ("What is the difference between bell fast in Chicago, Belfast in Maine and Belfast in Ireland?" "What is the Agynnian sect?"). Her proof, however, was not as it

---

* Isn't this once again the battle between taking what counts as belonging to Being or to Becoming, which we've seen fought out over mathematics and its objects?

appears in Loomis, who had clearly exercised his editorial powers over it (he *was* somewhat high in manner). Here is her proof rescued from the journal, with our reconstruction of what reasons weren't given there:

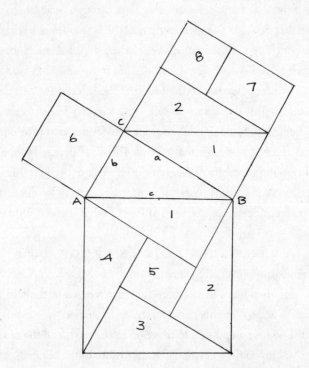

In the square on its hypotenuse, copy $\triangle ABC$ four times as shown, calling these copies 1, 2, 3, and 4. This leaves square 5, of side-length $a - b$, in the middle. Call the square on side b, 6.

Copy $\triangle ABC$ twice again in the square on the long side a, calling those copies 1 and 2. Divide the rectangle that remains in this square into a square, 8, congruent to 5, and rectangle 7.

Letting the numbers now stand for the areas of the figures they denote, Miss Coolidge needs to show that

$$1+2+6+7+8 = 1+2+3+4+5,$$

i.e., that

$$6+7+8=3+4+5.$$

Since 8 was constructed so that $8 \cong 5$, it remains only to show that $6+7=3+4$. But 3 and 4 each have area $\dfrac{ab}{2}$, hence $3+4=ab$; $6=b^2$ and 7 has sides b and $a-b$, hence area $b(a-b)=ab-b^2$.

So $6+7=b^2+ab-b^2=ab$, as desired.

What is particularly fascinating about this proof is that Miss Coolidge begins it geometrically—making and comparing shapes—but then turns to calculating with letters when, perhaps, she can't quite figure out what's left and what's wanting.

You might think that, being blind, Miss Coolidge would have resorted to abstraction as early as possible, but in fact she delays it as long as she can, trying to stay true to the spirit of geometry. She needs to account for square 5, so constructs the square 8 congruent to it in what remains of the square on a (after 1 and 2 are removed); and only then turns to a formula for calculating the area of that remaining rectangle, 7.

Had she not bothered to make 8, then the rectangle left in the square on a, after triangles 1 and 2 are removed, has area $a(a-b)=a^2-ab$, so that this rectangle, with 6, is a^2-ab+b^2; and that's just what areas $3+4+5$ add up to.

In fact, had she been wholly comfortable with symbolic manipulations, she could have avoided constructions in the square on a altogether, since $c^2=1+2+3+4+5=4\left(\dfrac{ab}{2}\right)+(a-b)^2=2ab+a^2-2ab+b^2=a^2+b^2$, the Pythagorean Theorem.

Having followed her proof, if we now follow her we may better understand the play between abstraction and different sorts of sensory information. But how find the woman behind Loomis's brief "Miss E. A. Coolidge"? It struck us that Coolidge is a good Boston name, and the journal in which her proof first appeared was published in Boston. Might she not then have been a student at the renowned Perkins School for the Blind? We e-mailed Jan Seymour-Ford, their research librarian, who answered: "That was inspired guesswork! Emma A. Coolidge was a Perkins student. She was born August 4, 1857, in Sturbridge, MA."

Emma had lost her sight from whooping cough when she was a year old—she could detect only light and shadow, and none of the remedies her parents tried (having her wear gold earrings, putting talcum powder into her eyes, blistering her temple with poisoned flies) helped. After graduating from Perkins, she studied at Wellesley for a year, then returned as a teacher to Perkins, where one of her students is said to have been Annie Sullivan, later Helen Keller's tutor. Emma married, had a daughter, wrote children's books, taught music in her New Hampshire village school, sewed, knitted, would kill and dress a fowl for dinner if her husband was away, and boldly went out alone, tying white rags to doorways so she could find her way from place to place.[4] Isn't this what we see in her proof: catching the chiaroscuro of prominent shapes, but navigating otherwise by those abstract relations with which practice and memory furnish the mind?

What sort of imagination this involves may matter here. The blind mathematician Louis Antoine was led by the famous analyst Henri

Lebesgue to study two- and three-dimensional topology, because "in such a study the eyes of the spirit and the habit of concentration will replace the lost vision." His equally outstanding compatriot, Bernard Morin, who has been completely blind since age six, was asked how he knew the correct sign in a long and difficult computation. "By feeling the weight of the thing", he said. More tellingly, Morin distinguishes between what he calls time-like and space-like mathematical imagination, and surprisingly says that he excels at the latter. A problem with picturing geometrical objects is that we tend to see only their outsides, which hide what might be complicated within. Morin, who works with extremely intricate objects in three dimensions, has taught himself how to pass from outside to inside (from one "room" to another). "Our spatial imagination", he says, "is framed by manipulating objects. You act on objects with your hands, not with your eyes. So being outside or inside is something that is really connected with your actions on objects."[5] Think of Emma sewing and knitting, or killing and dressing fowl.

Might Emma's extraneous calculations have come from an intrusion of the time-like into her sounder space-like imagination—and is the tactile yet one more intermediate between Being and Becoming?

THE HAWK AND THE RAT

You may not have dwelt on how Emma Coolidge packed her hypotenuse square—a way which could, we showed, by itself have proven the Pythagorean Theorem. In fact, unknown to her, it had—some twelve centuries before in India, by an astronomer and mathematician named Bhaskara. He speaks only of a 3-4-5 right triangle, and after a contorted verbal description deigns to say: "A field is sketched in order to convince the dull-minded", with this diagram:

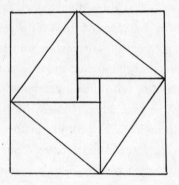

The problems he gives show that Bhaskara knew the Theorem in broader terms, although this pastoral example is still in the 3-4-5 family:

> A hawk was resting upon a wall whose height was twelve *hastas*. A departing rat was seen by that hawk at a distance of twenty-four *hastas* from the foot of the wall; and the hawk was seen by the rat [a feeling here for Nemesis as well as mathematics]. There, because of his fear, the rat ran with increasing speed towards his own residence which was in the wall. On the way he was killed by the hawk moving along the hypotenuse. In this case I wish to know what is the distance not attained by the rat, and what is the distance crossed by the hawk.*

---

* Too little given to tell? Bhaskara's problem supposes that wall and roaming ground are segments of intersecting chords of a circle, the latter a diameter; that the rat is slaughtered at the circle's center; and that the student knows that the products of segments of intersecting chords are equal. Hence in this figure, $144 = 12 \cdot 12 = y \cdot 24 \Rightarrow y = 6 \Rightarrow \text{diameter} = 30 \Rightarrow \text{radius} = 15$, so $x = 9$.

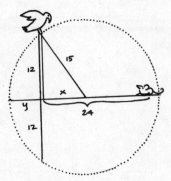

Life is not only stranger than we think, it may (to paraphrase J. B. S. Haldane) be stranger than we *can* think. Five hundred years later, another and much more famous Indian astronomer and mathematician named Bhaskara (no relation)[6] published the same figure (but flipped around the vertical axis and with the numbers and outside box removed) to accompany a proof of the general theorem.

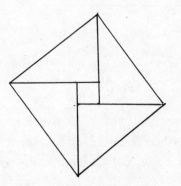

In the intervening centuries, however, the dull-minded had also been flipped into the sagacious, for this later Bhaskara explains the proof briefly, then says that just seeing the figure suffices. (In fact, after a terse summary, he writes: "And otherwise, when one has set down those parts of the figure there, seeing." Over the years this has been whittled down to the story that his proof consisted of no more than the figure and the single word, "Behold!")[7]

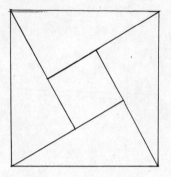

When you do, you see Guido's figure, built not on a square of side-length $a + b$ but of c, the hypotenuse. A negligible change, you think, until you try Guido-like rearrangements. You may find this one,

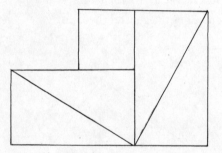

but it isn't nearly as pleasing as Guido's, since it needs a new line imagined or drawn in, others removed, and some mental arithmetic, to see the squares on the sides. And yet there is that powerful proof of the

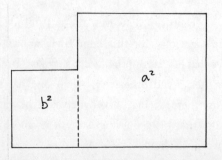

Theorem without any further drawing: the square's area, c^2, is made up of four right triangles, each of area $\dfrac{ab}{2}$, and that middle square, whose side-length is $b-a$. So

$$c^2 = 4\left(\frac{ab}{2}\right) + (b-a)^2 = 2ab + b^2 - 2ab + a^2 = b^2 + a^2,$$

as desired.

To achieve this result, we moved not shapes but symbols around: we construed rather than constructed—and from this slim difference opens the gap between geometry and algebra, between visual and abstract math.

Not that this is yet algebra, since its hallmark x, the unknown, has still to make its appearance. To understand, however, what a deep chasm has just been leapt, we need to sharpen our seeing of a split that is usually ignored. On the one side is thinking that uses placeholders—what was once accurately but misleadingly called "literal arithmetic" (so in $x+3=5$, x really means 'What?'; it holds the place for the answer '2', which immediately springs to mind). On the other side is algebra proper, with its unknown that itself later metamorphoses into a variable (in $x+3=y$, x and y vary together: if y were 5, x would be 2, but were y 13, x would be 10—that placeholder has become a variable). You see this distinction again in language, between names and nouns: to pass from bodies out there to names for them ('Adele' and 'Barbara' for the two girls whispering across the room, '3' for those sticks, | / \), is already a full turn of abstraction up our spiraling architectural instinct. To enmesh these names as nouns in a grammatical matrix is a further turn, freeing them now to move together into unguessed configurations. A third, algebraic, turn will invite us to solve for what is still missing, just by grammatical rearrangements of our nouns ($3 + x^2 = 19$—ah, $x^2 = 16$, so x is 4 or -4). How do these changes happen—how do we think our way from $3^2 + 4^2 = 5^2$ and $5^2 + 12^2 = 13^2$ to the universal $a^2 + b^2 = c^2$? Take the simpler case of area. We replace specific numbers by generic names (4 becomes the triangle's 'base' and 3 its 'height') and describe its area no longer as $3 \cdot \dfrac{4}{2}$ but as 'half the base times the height'.

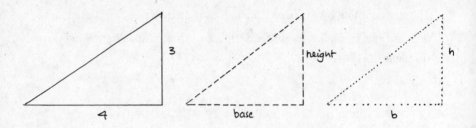

Then, by what seems a trivial act of abbreviation but in fact resounds with consequences, we write: 'A', '*b*', and '*h*'. We next clamp these symbols (somewhere between nouns and pronouns) together in a 'formula'—a little form:

$$A = \frac{bh}{2}.$$

Only now can we hear and respond one further turn up to algebra's call: if you knew that A was 6 and *b* was 3, what must *h* be? Grammar ignores the objects (in this case shapes) its nouns stand for and allows us to rearrange, wholly within this new *linguistic* context: $h = \dfrac{2A}{b} = 2 \cdot \dfrac{6}{3} = 4.$

Haven't we just found the proper context in which to see, as Parmenides did, that motion is an illusion and only Being is? Euclid constructed temporary scaffoldings. Later (in the proof that involved tiling the endless bathhouse floor), invoking infinity allowed us to keep the objects of mathematics still, while godlike thought hurried over them. Here those objects are if anything yet more at rest as abstraction grows, shape yields to form, and movement is confined to symbols in our language (or in our thought, if language faithfully mirrors it—or in both, as is the way with reflection).

Bhaskara's diagram needed none of this third-storey work, but the grammar of these symbols (aptly called *rules* rather than *laws* of multiplication, subtraction, and distribution) sufficed on the second level to rewrite $(b-a)^2$ as $b^2 - 2ab + a^2$, which, along with grammatical steps on the same level, produced the Pythagorean Theorem from this orderly dance of signs. We are centuries past diagrams in the dust.

Are these what you said were no more than tautologies, Russell? Meaning is the aftermath of form. We want to see the hang of things, and abstraction's upward path leads us, like Dante's, to it. Yes, but at what cost? What have you relished more in this book so far, the glimpses of Hippasus and Miss Coolidge, of Loomis and Thabit, or the growing clarity of a once hidden harmonious structure? We can have both. The signs that decorate this structure pick out doors opening inward on the vast tower of time. Behind each one are singular people, odd events, histories of invention that led to discovery, and the marks that personalities have left on impersonal truth—which give even it a singular character too.

△ △ △

What would you do were you carrying a precious platter, and tripped? Were you frivolous and French, and a footman in the service of the Sun King, Louis XIV, you would sweep the shards aside while an even more glorious platter, bearing an even more sumptuous bird, was brought in by the footman waiting in the kitchen for this planned accident of conspicuous consumption. Were you Greek, serious, and talking with Socrates, you would agree that the One had thus lightly become the Many.[8] But were you Chinese and living a millennium ago, you would (the legend runs) have gathered up the fragments in a panic and tried to reassemble them into their original square shape—and failing, would have instead devised the thousand patterns of the Tangram: the puzzle that once rapt people away as fully as Sudoku does now.

The pieces, called tans, have sedately settled into seven fixed shapes: a pair of small and a pair of large isosceles right triangles; one more, of middle size; a square; and a disconcerting parallelogram—all made to pack into an attractive square box.

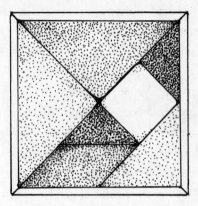

The game is to rearrange them all to make the countless different figures whose outlines only are in the booklet that came with your puzzle: a cat, say,

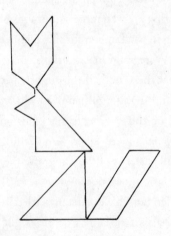

or a swan:

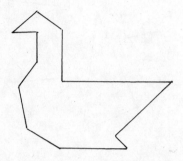

Shall we see if the Pythagorean Theorem is among them?

Here are the pieces in their box, making the square on the hypotenuse, and (if our given right triangle is isosceles), taken out of the box and arranged to make the squares on the two sides:

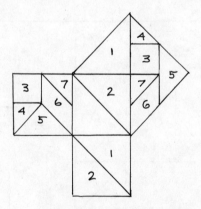

You may feel that the arbitrary number and shapes of a Tangram make this not so much a problem as a puzzle (is this another criterion to add to our classification of proofs—two genera that will each branch out from the families, with the hint of more finely twigged species in between?). And you may find it still more disappointing that we can't generalize with tangrams to other right triangles. Well, were we not bound by historical convention, we could make these instead be our seven pieces:

For now they will miraculously do just what we had wished (once again, showing them packed and unpacked around the unnumbered right triangle):

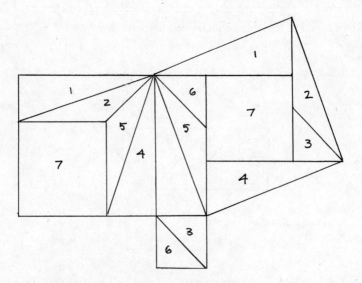

Even more of a miracle—Behold! Isn't this just Bhaskara's figure? But with those two cuts giving us seven pieces from his original five, we can now do what he couldn't and make the small and larger square without either invisible lines or mental arithmetic (we've shown here, in one picture, the seven pieces arranged to make the square on the hypotenuse, and, differently arranged, to make the squares on the two sides of the un-numbered right triangle).

This idea came, twenty-five years ago, from the Orientalist Donald Wagner,[9] who later, however, found out that he had been preceded by a young German mathematician, Benjir von Gutheil, killed in the trenches of France in 1914.

If we look past the legend of the Tangram's origin, and its slightly more probable source in furniture sets of the Song Dynasty, we may come on Liu Hui's third-century A.D. commentary on a text two or three centuries earlier, *The Nine Chapters on Mathematical Art*. The Pythagorean Theorem, for a 3-4-5 right triangle, plays a large part in it, and although none of the diagrams that were probably in Liu Hui's commentary have survived, here is a playful attempt to reconstruct them.[10] Like one of those antique inlaid puzzle boxes, it is made with sliding panels and hidden pressure-points.

Given that 3 and 4 were important numbers in ancient China, since 3 was taken to be the circumference of a unit circle while 4 was the perimeter of a unit square,[11] it seemed reasonable to look for a circle and square inside a 3-4-5 right triangle—and this is what D. G. Rogers did. The Tangram-like interpretation of Liu Hui that he came up with in fact generalizes to any right triangle.

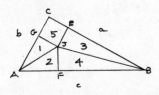

In right triangle ABC, bisect angles A and B, and let these bisectors meet at J (the center of the triangle's incircle—the circle, that is, tangent to the triangle's three sides—our hidden pressure-point).

Drop perpendiculars from J to the sides, meeting BC at E, AB at F, and AC at G. Number the resulting triangles as shown.

$\Delta 1 \cong \Delta 2$, $\Delta 3 \cong \Delta 4$ (by shared sides, right angles and the equal, corresponding, bisected angles). Hence $JG = JF = JE = r$, the incircle's radius. Let $2r = d$, this circle's diameter.

You see that $JG = JE$ makes 5 a square.

Taking 1, 2, 3, 4, and 5 as our tans, slide these panels like this:

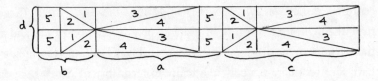

Make three more copies of this tangram and put all four together:

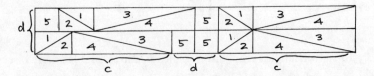

Note the lengths a, b, and c, as shown; call the height of this rectangle d.

The area of this inlaid rectangle is $d(a+b+c)$, but since each of the four blocks composing it has the area of the original triangle, namely $\frac{ab}{2}$, we have

$$d(a + b + c) = 4\frac{(ab)}{2} = 2ab, \text{ hence } d = \frac{2ab}{a+b+c}. \tag{1}$$

We want to unlock this box, so, sliding the panels in the lower left rectangle only, we have

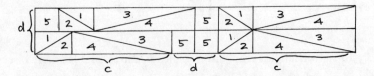

Comparing the lower edges of our two four-block formations, we see that

$$a + b = c + d, \text{ so that } d = a + b - c \tag{2}$$

Equating (1) and (2),

$$\frac{2ab}{a+b+c} = a + b - c \Rightarrow 2ab = (a + b + c)(a + b - c)$$
$$\Rightarrow a^2 + b^2 = c^2.$$

Are you giddy with all this prestidigitation, or is that a sudden feeling of déjà vu? Didn't we see something very like this back in Loomis's proof of February 23, 1926? There he used only the bisector of ∠A and the perpendicular to AB, from the intersection of the bisector with BC, and finished his proof with no more than three tans and a fourth of the steps. And we thought that was baroque!

But if you want déjà vu all over again, isn't this also Guido's proof, with an intaglio of extra lines? Rogers himself rearranges his twenty tans in those tried and true ways:

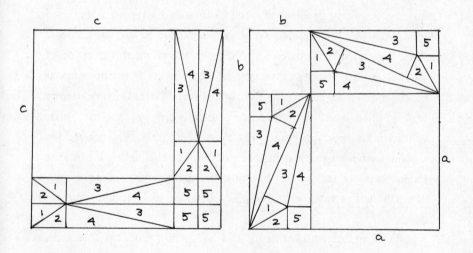

We may not have to push our invention quite this hard in order to reconstruct what the Chinese mathematicians of the third century A.D. had in mind. A 1213 edition has survived of a commentary by Liu Hui's contemporary Zhao Shuang on *The Gnomon of Zhou* (a book even earlier than *The Nine Chapters*), with this "Figure of the Hypotenuse",[12] its four vermilion triangles surrounding the solitary central yellow square. Look familiar?

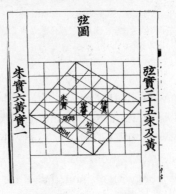

With all sorts of communication over the centuries having been possible between the Far East and India, would you want us now to revise our attribution of this figure from Bhaskara to *The Gnomon of Zhou*? There is the slight problem that the Chinese stuck resolutely to 3-4-5 triangles, while Bhaskara, by avoiding specific numbers, made his proof work for any right triangle at all. More to the point, this question of chest-thumping priority detracts from what counts in mathematics as much here as it did among the Pythagoreans and Greeks. Every attempt to shore up a proof by invoking remote authority weakens the innate validity of its impersonal argument. Add, besides, the evidence of Miss Coolidge having given parthenogenetic birth to this figure: it can arise independently, and doubtless has, time and again, being one of the characters in the abstract repertory company that performs our inner commedia dell'arte.

Einstein, after all, came up as a boy with a proof via that other mask, Pythagorean proportions:

> I remember that an uncle told me the Pythagorean theorem before the holy geometry booklet had come into my hands. After much effort I succeeded in "proving" this theorem on the basis of the similarity of triangles ... for anyone who experiences [these feelings] for the first time, it is marvelous enough that man is capable at

all to reach such a degree of certainty and purity in pure thinking as the Greeks showed us for the first time to be possible in geometry.[13]

Does it take anything away from his discovery that others had made it before him? Georg Christoph Lichtenberg once wrote: "What you have been obliged to discover by yourself leaves a path in your mind which you can use again when the need arises." As, for Einstein, it did, and he did.

Of much greater significance in this Chinese "figure of the hypotenuse" is what it tells us about the risks with which we use examples. When we point, we want others to look where we're pointing, not at the end of our finger. One problem is the nervous viewer's eagerness to take us exactly at our word—but another is the speaker's failure to imagine how many types of ambiguity he is open to. Drawing these 3-4-5 right triangles in a 7×7 grid might have been meant to stand for any right triangles, with the grid no more than a concession to artistic taste, architectural convention, and printers' convenience.

What comes across from text and commentary, however, is that the grid was essential, and that the diagram meant exactly what it shows: the Pythagorean relation *belongs* to this single case, which is also singular, given the mathematical importance of 3 and 4.* This is no mere example but The Exemplar. It isn't meant to conjure up differently proportioned right triangles—only perhaps to remind us that 3, 4, and 5 model the order of the world, and that balance within each family, in all the vermilion land, is guaranteed by the Emperor, in his central Yellow Palace, having the mandate of heaven.

Our common humanity makes it hard to imagine that another society might have weighted its emphases so differently from ours as to think that looking at an isolated instance, the 3-4-5 right triangle, showed what

---

* It was important as well that this was a *right* triangle: for the Chinese seem to have thought of other triangles as more or less filled with area, but of a right triangle as a framework whose base and height *determined* its area (thus serving as the foundation for surveying the heavens and earth).

was true for all—and that this "all" wasn't other instances but algorithms based on manipulations on and with it. The diagram *instantiated* the algorithms. "The key fact", says the historian Karine Chemla, "is that these figures form the basis for the whole development in the following sense: all the algorithms placed after the figures derive from them, in that the reasons for their correctness are drawn from these figures and only from them."[14]*

How to embody the general accurately yet suggestively: that was the great leap from the specific diagrams of the Babylonians to the unmarked figures of the Greeks. While exemplary thought may promote a spread of relations from those given, it is justified in broader contexts only by the validity of these relations in the narrow original.

We turned away from Liu Hui because Zhao Shuang's commentary seemed to give us more of an insight into the thought of the time—but it would be wrong to sell short a scholar of Liu's intelligence. He worked within a tradition so respectful of its ancients that correcting them wasn't acceptable. Liu may well have understood that the structure of this proof supported its generalization to all right triangles, but would have had to express himself in written or spoken asides.[15]

△ △ △

Loomis interrupts the march of his crabbed diagrams suddenly, after his Geometric Proof 52, which bristles like a hedgehog:

---

* This is, after all, an understandable direction that figures might take, from storing, supporting, and enhancing structural arguments to replacing them.

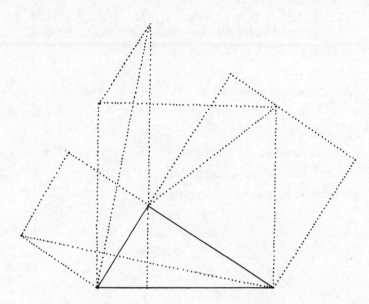

He writes:

This figure and proof is taken from the following work, now in my library, the title page of which is:

Euclides Elementorum Geometricorum

Libros Tredecim

Isidorum et Hypsiclem

& Recentiores de Corporibus Regularibus, &

Proclii

Propositiones Geometricas

The work from which the above is taken is a book of 620 pages, 8 inches by 12 inches, bound in vellum, and, though printed in A.D. 1645, is well preserved. It once had a place in the Sunderland Library, Blenheim Palace, England, as the book plate shows—on the book plate is printed—"From the Sunderland Library, Blenheim Palace, purchased, April, 1882." I found the book in a second-hand book store in Toronto, Canada, and on July 15, 1891, I purchased it. E. S. Loomis.

The work has 408 diagrams, or geometric figures, is entirely in Latin, and highly embellished.

Ah, Loomis, Loomis.

△ △ △

Take that tumble of right triangles you saw in Chapter Three (p. 55). What if each had been scaled up from the first (which has sides *a*, *b*, and *c*)—one of them by *a*, one by *b*, and one by *c*:

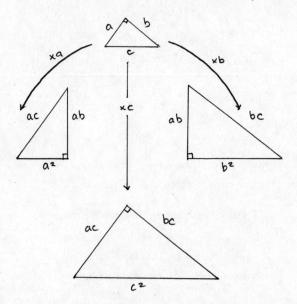

Since the first and the second now each have a side *ab*, paste those sides together to form a new right triangle with legs *ac* and *bc*, and hypotenuse $a^2 + b^2$. But the third right triangle, scaled up by *c*, also has legs *ac* and *bc*, with hypotenuse c^2. Since these two triangles are congruent (by S.A.S.), their hypotenuses are equal, so $a^2 + b^2 = c^2$.

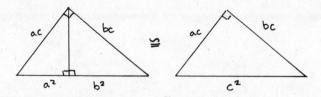

A miraculous proof.[16] Once again, construing a mute figure in two different ways makes it speak. For all the translucency of Guido's and the old Pythagorean inward construction, is this outward expansion instead the most transparent of all?

△ △ △

By contrast, we came on a proof in a professorial note titled "Pythagoras Made Difficult",[17] with the writer's reflection that "If you are feeling particularly hostile toward the whole universe and you want to do something evil, show this argument to your calculus students and tell them they need to learn differential equations to understand how to prove the Pythagorean theorem. With any luck, half of them will believe you."

Whether or not you were Professor Hardy's student, if you are no friend of differential equations, heed the epitaph W. B. Yeats wrote for himself: Horseman, Pass By!

Otherwise, consider this diagram with the familiar beginning of the alphabet replaced by the variables of calculus, x, y, and z.

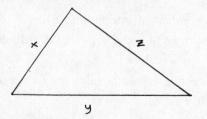

Look at a very small triangle similar—with a flip—to the whole, tucked into its left-hand corner (this triangle is cut off with a stroke perpendicular to y):

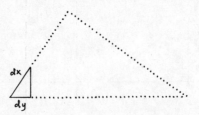

Its side-lengths, in calculus-speak, are *dx* and *dy*: increments that will tellingly dwindle to nothing. The similarity of the small to the large triangle gives us (short/long):

$$\frac{dy}{dx} = \frac{x}{y}.$$

Using the powerful technique called "separation of variables" (miracled up in the seventeenth century by Guillaume François Antoine, le Marquis de l'Hôpital), we can rewrite this equation as

$$y \, dy = x \, dx$$

and integrate:

$$y^2 = x^2 + K,$$

where K is the "constant of integration". Being a constant, its value doesn't change as those of *x* and *y* shrink, so that, at the limit when $x = 0$, $y^2 = K$.

But look at the diagram: as the side *x* goes to 0, *y* collapses onto *z*— that is, when $x = 0$, $y = z$, hence $y^2 = z^2$.

This means that the constant K is now and always has been z^2, so

$$y^2 = x^2 + z^2.$$

"There now, wasn't that easy?" as the Contortionist said to the Fat Lady.

HILBERT IN PARIS

What we gain in new proofs we lose in attempts to classify them, trawling with anti-torpedo nets for minnows. Perhaps, though, we needn't carry rigor as far as mortis in order to satisfy our legal longings and understand better what we want of a proof. Tying an impersonal truth to the axioms of our making?* Certainly. Doing this gracefully, and with the sudden surprises that mark revelation? Yes—so that to Bhaskara's "Behold!" the Miss Coolidge we shelter answers, "I see!"

In 1900 David Hilbert announced twenty-three problems at the International Congress of Mathematicians in Paris. These picked out the boundaries of mathematical inquiry for the century ahead (ten have since been solved, five not; the rest swelter in various kinds of limbo). A twenty-fourth, which never made it past his journal, asks for a way of determining the simplest proof of a theorem (he cites the Pythagorean)—as by counting the number of steps needed, once the chain of reasoning has been methodically reduced.[18]

By asking for the simplest, and asking in these terms, Hilbert confronted the formal and the intuitive. But are simplest and shortest really the same? Hasn't many a shining explication of complexity emerged from mazy wanderings, and elegance, as in music, from prolonging a single note into a trill? Would we have noticed the profundity of $3^2 + 4^2 = 5^2$ had it dawned on us as negligently as $3 + 4 = 7$?

The question of simplicity isn't itself simple. It is inextricably bound up with the vertiginous issues of consistency and completeness, and of judging whether the truth of a mathematical statement can be decided mechanically. These are fraught issues, since the axiomatic foundations of the rigorous formal methods for laying bare the armature of such statements can, ironically, themselves be no more than *intuitively* convincing (or should their fecundity, say, rather than their clarity, be the criterion for choosing these axioms?). Should every step along the way

---

* Plotinus somewhere sets up a proportion: the immortal soul is to the body as the truth of a theorem is to the proof cobbled together for establishing it.

have a finite character, or must "ideal" statements (i.e., statements about infinite collections) be added to them? Simplicity is the very devil.*

Categorizing may cause what is ultimately a single proof, pure as a diamond, to seem like a multitude, as we are dazzled by the shimmer reflecting off its facets. Not only does this hide from us a One behind the Many, it keeps us from seeing the endless subtle issues at play in the background, such as (with the Pythagorean Theorem) the nature of area and the architecture of motions on the Euclidean plane, and in general the opposite pulls of abstraction and embodiment.

So if we are not to categorize proofs of the Pythagorean Theorem by some scheme or other, nor dare to arrange them by degrees of simplicity, aren't we led to cultivating our taste for them in terms of what light they shed on their locales, whose shadowed depths in turn develop them?

△ △ △

The Greeks, we said, believed that a rare mortal might touch the bronze sky for a brief moment. Twenty-five hundred years ago the Cretan runner Ergoteles won the long race at the Olympic Games, and was the glory of his time. In 1924 Mr. Ericson, the math teacher at Milwaukee's Washington High School, told his junior class that if any of them could work out a new proof for the Pythagorean Theorem, he would see to it "that the matter would be properly advertised in the local newspaper." After a time Alvin Knoerr came forward.

---

* Or was Hilbert right in thinking (like Leibniz before and Einstein after him) that the maximum of simplicity and perfection is realized in the universe? See Thiele, op. cit., 18.

"Although I never expected to be able to develop a new proof," Knoerr later wrote,[19]

I tried a little experimenting a few days later just to pass time. This proved so interesting that I continued working out all kinds of ideas. The usual result of fifteen or twenty minutes of work, however, was that the solutions would boil down to an identity, thus bringing me back to the original starting point.

After experimenting in this manner for about a month I began to realize that my efforts were useless. Finally I struck upon a plan which seemed to me to be far more logical than others because instead of starting from the beginning I started from the end and took the fact that the square on the hypotenuse is equal to the sum of the squares on the other two sides for granted.

Then I broke this equation up into several other equations and finally ended up in a proportion. We had just finished similar triangles so the proportion which I had developed suggested the usage of similar triangles. Then after I had the triangles constructed it was a matter of only an hour or so to work out the proof.

Here is what Alvin Knoerr came up with, simplified a bit from his Rube Goldberg original.

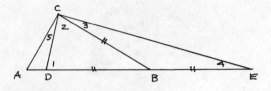

Given: $\triangle ACB$ has a right angle at C.

AB is extended to E, so that $CB \cong BE$;

point D is placed so $DB \cong CB \cong BE$.

1. In $\triangle CAD$ and $\triangle CAE$, $\angle CAD \cong \angle CAE$.
2. $\triangle CDB$ is isosceles, so $\angle 2 \cong \angle 1$.
3. $\triangle CBE$ is isosceles, so $\angle 3 \cong \angle 4$.
4. By addition, $\angle 2 + \angle 3 = \angle 1 + \angle 4$.
5. In $\triangle DCE$, $\angle 2 + \angle 3 + \angle 4 + \angle 1 = 180°$.
6. So $\angle 2 + \angle 3 = 90°$—i.e., $\angle DCE = 90°$.
7. In $\triangle ACE$, $\angle 5 = \angle 5 + \angle 2 + \angle 3 - 90°$.
8. In $\triangle ACE$, $\angle 3 = \angle 5 + \angle 2 + \angle 3 - 90°$.
9. Therefore, $\angle 5 \cong \angle 3$.
10. And, by transitivity from step 3, $\angle 5 \cong \angle 4$.
11. Therefore, $\triangle CAD \sim \triangle EAC$: $\angle CAD \cong \angle EAC$, $\angle ACD \cong \angle AEC$ and $\angle CDA \cong \angle ECA$.
12. So $\dfrac{AE}{AC} = \dfrac{AC}{AD}$.
13. But since $AE \cong AB + CB \; (\cong BE)$
     and $AD \cong AB - CB \; (\cong DB)$,
14. then, substituting in Step 12, $\dfrac{(AB + CB)}{AC} = \dfrac{AC}{(AB - CB)}$.
15. Multiplying: $(AB + CB)(AB - CB) = AC^2$
16. $AB^2 - CB^2 = AC^2$
17. So $AB^2 = CB^2 + AC^2$.

Alvin Knoerr's reminiscence is the only record we have found of the creative work behind a proof of the Pythagorean Theorem. How representative of others was the movement of his thought back from conclusion to premises, then forward again—the method often called "analysis and synthesis"? Removing the scaffolding before unveiling the building heightens the jolt that joins the true to the beautiful, but loses us the contemplative pleasures of watching the mind shape and mortar its bricks.

What became of Ergoteles after he won his town the race? He was granted citizenship in the Sicilian city of Himera, where he had fled after an uprising in Cretan Knossos, and later gained the coveted right to own land. And Alvin Knoerr? We catch a glimpse of him in 1936, tramping out a gigantic *M* on a snowy hillside in Wisconsin, to prove that the projected monument to the School of Mines, where he was a student, would be visible from distant Platteville. Is that him, prospecting for atomic minerals after the war? He edited the *Engineering and Mining Journal* in the 'sixties, writing that most research "is nothing more than searching for ideas that have already been set down in print." He died, eighty-eight years old, in 1995, in Queens, New York. His hometown historian could find no traces of him. He has even been deleted from Wikipedia.*

△ △ △

Here is a proof with a hypermodern flavor, outflanking the Theorem. It, too, cunningly flirts with, then slips past, trigonometry.

A right triangle is uniquely determined by the length of one side and the size of an acute angle. Let's choose *c* as our side and the smallest

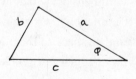

angle, ϕ. We know this triangle's area is

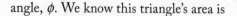

<hr />

* Not only Milwaukee's historian but the offices of VISIT Milwaukee, and of Public Relations, and a local journalist, came up empty-handed. It took the inspired research of Amanda and Dean Serenevy to find articles in obscure journals and relevant years of *The Scroll*, the Washington High School yearbook—where Alvin is described as "Tall, straight, and thoroughly a man, a fine example of an American."

$$A_\triangle = \frac{1}{2}\,\text{base}\cdot\text{height} = \frac{1}{2}c\,\sin\phi\cdot c\,\cos\phi = c^2 f(\phi),$$

where the function $f(\phi)$ in this case is $\dfrac{\sin\phi\cdot\cos\phi}{2}$.

What we'll do now is copy $\angle\phi$ in the right angle, giving us triangles 1 and 2:

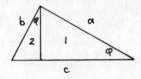

Let's look at areas again:

$$A_{\triangle_1} = a^2 f(\phi),\ A_{\triangle_2} = b^2 f(\phi),$$

so since $A_{\triangle_1} + A_{\triangle_2} = A_\triangle$,

$$a^2 f(\phi) + b^2 f(\phi) = c^2 f(\phi).$$

Dividing by $f(\phi)$ gives us $a^2 + b^2 = c^2$.

What happened? The pea must have been under one of those shells, and now it's gone!

This is an instance of 'dimensional analysis', where we needn't inquire into the nature of the function f, just glean the desired result from this way of looking, which dates back to the physicist Percy Bridgman in the 1920s.[20] It allows you to solve a problem by attending to no more than the dimensions of its variables (thus finding easily, for example, that the drop in pressure of a fluid flowing through a pipe is proportional to Q/R^4, where R is the pipe's radius and Q the volume of flow in a unit of time). You might think of this proof as a streamlined version of the Pythagorean, sped up by removing all the finicking Euclidean constrictions.

$\triangle\ \triangle\ \triangle$

Having crossed the divide from mathematics to physics, let's linger a moment to relish a proof that depends on the impossibility of perpetual motion. Make yourself a box with parallel congruent right triangles ABC for top and bottom, and with side height h. Fill it with gas at pressure p, seal it closed, then hinge it at corner A to a vertical pole, around which it can pivot freely.

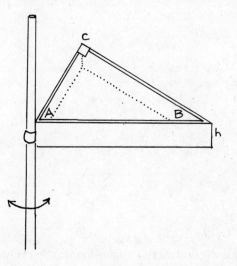

We define *torque*, T, as the twisting ability of a force F on a lever of length k about a point A: $T = F \cdot k$. These levers will be the faces of our box. As good Newtonians, we make these assumptions:

1. There is no perpetual motion.
2. A gas pushes equally on all the faces of its container.
3. The force F with which the gas pushes on a face is concentrated at the midpoint of that face.
4. This force is the product of the gas pressure, p, times the area of that face.

Let's now calculate these forces. We'll ignore those acting on the top and bottom faces (those right triangles), since they are equal and opposite, hence cancel each other out.

The force acting on the AB face, with area $h \cdot AB$, is

$$F_{AB} = p \cdot h \cdot AB.$$

As you see from the diagram, this force will act to rotate the box clockwise.

The forces acting on faces AC and BC try to rotate the box counterclockwise.

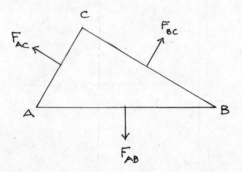

They are: $F_{AC} = p \cdot h \cdot AC$ and $F_{BC} = p \cdot h \cdot BC$.

We can now calculate the torque about A. Since we're taking the force to act on each lever's midpoint,

$$T_{AB} = F_{AB} \cdot \frac{AB}{2} = p \cdot h \cdot AB \cdot \frac{AB}{2} = p \cdot h \cdot \frac{AB^2}{2}$$

$$T_{AC} = F_{AC} \cdot \frac{AC}{2} = p \cdot h \cdot AC \cdot \frac{AC}{2} = p \cdot h \cdot \frac{AC^2}{2}$$

Finally, since C is a right angle, we can in our minds slide BC down along AC to A without changing its angle, so that the force F_{CB}, acting on CB's midpoint, indeed gives us the torque T_{CB} around A:

$$T_{CB} = F_{CB} \cdot \frac{CB}{2} = p \cdot h \cdot CB \cdot \frac{CB}{2} = p \cdot h \cdot \frac{CB^2}{2}.$$

Now here is the crux of the argument. Because we assume that there can be no perpetual motion, the clockwise and counterclockwise torques must balance:

$$T_{AB} = T_{AC} + T_{CB}$$

That is,

$$p \cdot h \cdot \frac{AB^2}{2} = p \cdot h \cdot \frac{AC^2}{2} + p \cdot h \cdot \frac{CB^2}{2},$$

and simplifying, $AB^2 = AC^2 + CB^2$, as we had hoped.

Toto, where are we? Very far from the fruited plains of mathematics. All that talk of circular motion may stir up fears of a deeper circularity: not that the Pythagorean Theorem might already lie in the handling of vectors here, but that the axioms of physics invoked may reflect an underlying geometry, as if we were deriving this proof not from axioms at all but from their consequences. That would make the result not so much a proof as an illustration—a drawing-down of abstract structure into the visible. And what exactly is the turn-and-turn-about relation between physics and mathematics: do the insights of each propel the other? Are they hopelessly or hopefully entangled? May an anti-cyclone whirl us home again.[21]

LE VASISTDAS

A curious French name for a small window that can be cranked open, like a transom, is *le vasistdas*, from the German *Was ist das?*, "What is

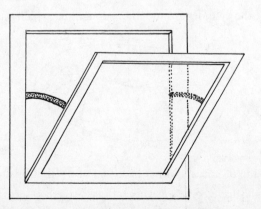

that?" (claimed to be what provincial German soldiers said when they first saw these transoms, in 1870s Paris). You might want to call *this* diagram a vasistdas:

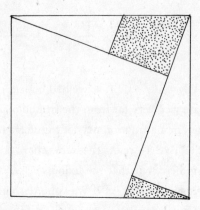

What does it look out on, with its strangely leaded lights? Well, take this square by its top edge and bend it down and around into a cylinder, gluing top to bottom:

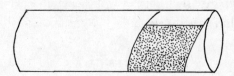

Next take what were its left- and right-hand sides—now the left and right circles at the ends of the cylinder—and glue them together, having bent the cylinder around into a doughnut (which has the

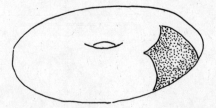

glorified name of 'torus'). What have you now? That quadrilateral and those triangles have come together into two squares, so that the initial square (which you could take as having been that on the hypotenuse) is fully covered by the squares on the two legs! This is the other world that window disclosed, the farthest fetch of a proof, it seems, by way of mapping the parts onto a torus made of their whole.[22] When the Cheshire cat disappeared, at least it left its smile behind. Here the original right triangle has vanished into the doughnut's hole.

△ △ △

We've sailed past so many islanded proofs—stopping to dine on some and waving at others from afar—that we begin to feel as legendary as Odysseus. Is that Samos over there, where Pythagoras was born? Kepler thought that the soul of Pythagoras might have migrated into his own. But are all these proofs, traveling abroad from the inland sea, not his true reincarnations?

CHAPTER SIX

Exuberant Life

The Pythagorean Theorem is an ancient oak in the landscape of thought. We've traced many of its roots down to where they clutch the rocks of our intuition—those proofs that in fact make it a theorem. But trees have their branches in the air, growing more luxuriantly in the direction of light and open spaces, more densely toward neighboring woods, creating the local chaos and changing symmetry that mark living things.

The tree of man was never quiet, nor are the trees in the forest of his certainty, taking the least opportunity to broaden and so conquer more of thinking's space. Here we'll follow some striking ways that the Pythagorean Theorem has spread out. First, though, since generalizing may precede as well as follow, let's look at how the Pythagorean relation can be drawn from one yet broader.

Its source is Ptolemy, whom you met in Chapter Two with his wonderful method of approximating square roots by gnomons. He was a Roman citizen in the second century A.D., a member of Alexandria's lively Greek community, steeped in Babylonian data, and called in later Arabic sources "the Upper Egyptian"—more a confluence than a person. His *Almagest* (the Arabic rendering of the Greek *Syntaxis* as "The Great Work") established a geocentric model for astronomy on the basis of detailed mathematical research—which included what came to be called Ptolemy's Theorem.

For Ptolemy saw that if a quadrilateral ABCD were to be inscribed in a circle,

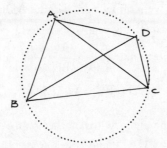

along with its diagonals, then the product of (the lengths of) these diagonals would equal the sum of the products of the opposite sides:

$$BD \cdot AC = AB \cdot CD + AD \cdot BC.$$

This hardly leaps off the page at you, so that just seeing it is already worth a niche in the mathematical Pantheon. Much less obvious is how to prove this—which Ptolemy did with astonishing economy (and this leads the sun to shine on his bust). He simply drew a line from A to BD, meeting it at E, making $\angle BAE \cong \angle CAD$.

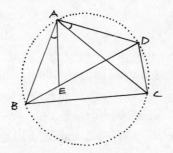

He then went to work with similar triangles, the ratios of their sides, and the fact that if two angles inscribed in a circle cut off the same arc, they are equal.

This is how he did it.

1. Adding $\angle EAC$ to the equal angles $\angle BAE$ and $\angle CAD$ means that $\angle BAC \cong \angle EAD$.

2. Since arc AB is cut off by $\angle ACB$ and $\angle ADB$, $\angle ACB \cong \angle ADB$.

3. Two pairs of equal angles give us $\triangle EAD \sim \triangle BAC$, hence their sides are proportional: $\dfrac{AD}{ED} = \dfrac{AC}{BC}$ —or multiplying out,

$$AD \cdot BC = AC \cdot ED.$$

4. Play this game again with another pair of angles: we already have $\angle BAE \cong \angle CAD$, and (since arc AD is cut off by $\angle ABE$ and $\angle ACD$) $\angle ABE \cong \angle ACD$.

5. These two pairs of equal angles give us $\triangle BAE \sim \triangle CAD$, hence $\dfrac{AB}{BE} = \dfrac{AC}{CD}$ —that is,

$$AB \cdot CD = AC \cdot BE.$$

6. Now notice what Ptolemy had in mind all along: $BD = BE + ED$.

7. So the product we want, $BD \cdot AC$, is equal to $(BE + ED) \cdot AC$, hence $BD \cdot AC = BE \cdot AC + ED \cdot AC$, which (thanks to steps 3 and 5)

$$= AB \cdot CD + AD \cdot BC.$$

Shall we say: Q. F. P.

—quod fecit Ptolemaeus?*

Ptolemy needed this theorem to deduce the key trigonometric formulae for sines and cosines of angle sums and differences—but he brought someone else home as well from his circle dance, since the Pythagorean Theorem follows when the inscribed quadrilateral is a rectangle: for now $AB = CD$, $AD = BC$, and $AC = BD$, so

$$AB^2 + BC^2 = AC^2.$$

---

* Generalizations flow through an endless array of nested cones. This theorem, for example, follows easily in a remote *inversive geometry*, where the ingenuity Ptolemy came up with is quite differently stored in the initial idea of reflections in a circular mirror (which turns his cyclic quadrilateral into line segments). For a proof done in this way, see Tristan Needham's *Visual Complex Analysis* (Oxford, 1998), pp. 138–9.

SHAPES OTHER THAN SQUARES

In Chapter Four you saw how Euclid extended the Theorem from squares to any similar polygons on a right triangle's three sides. More than a century before him, Hippocrates of Chios was thinking about semicircles on those three sides—for these too will be similar (all semicircles are). He used this fact in what looks like an attempt to square the circle.

It was he who, in Chapter Three, came up with two mean proportionals in hopes of doubling the cube. The Greeks were equally troubled by how to make (with straightedge and compass) a circle that had the same area as a given square, and his idea may have been that forming a controlled part of a circle with the same area as a right triangle would have been a sizable stride along the way. The part he found was a lunule, a quarter moon—

and while this didn't turn out to help with squaring the circle (nothing would have: it took twenty-two hundred years to prove that the circle couldn't be squared), it did leave us this attractive consequence of the Pythagorean Theorem:

THEOREM: On right triangle ABC construct semicircles on sides AC and BC, and inwardly on AB, giving us the regions labeled Q and R in this diagram. Then $|\Delta ABC| = |Q| + |R|$.

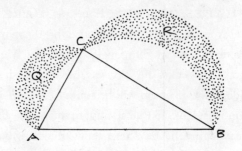

PROOF:

1. By the Pythagorean Theorem we know that $AC^2 + BC^2 = AB^2$.

2. So of course $\left(\dfrac{AC}{2}\right)^2 + \left(\dfrac{BC}{2}\right)^2 = \left(\dfrac{AB}{2}\right)^2$. Take these as the radii, r_1, r_2, r_3 of circles on those three legs,

3. and, multiplying through by $\dfrac{\pi}{2}$,

$$\frac{\pi}{2}r_1^2 + \frac{\pi}{2}r_2^2 = \frac{\pi}{2}r_3^2.$$

That is, the area of the semicircle on AC and the area of the semicircle on BC equals the area of the semicircle on AB— even when you see that last semicircle on AB as folded back over the triangle.

4. Lettering the parts as in this diagram,

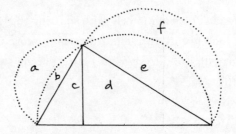

$$a + b + e + f = b + c + d + e.$$

5. And canceling like terms, $a + f = c + d$.

That is: the two lunules Q and R have their areas add up to that of $\triangle ABC$.

△ △ △

Squares, polygons, semicircles—aren't we about to swing open the shutters again? For as long as a figure has area, that of those similar to it will vary as the *square* of corresponding linear measures—in our case, the line-segments attached to our triangle's three sides. What you're seeing is the continuing abstraction from 'square' as denoting a shape, to connoting the measure of area per se. This process was under way several chapters back, when we spoke of a shape not *being* but *having* a square.

Read in this way, the Pythagorean Theorem therefore holds even when the most art nouveau shapes flourish on a right triangle's hypotenuse, along with shapes similar to it on the legs. They can, if you wish, be as lacy as your great-grandmother's antimacassars, so long as they have areas:

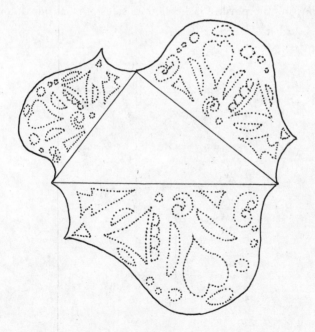

The wonders of calculus allow us to determine the areas of such lacework (the integral began its life with just this aim), by extending the notion of triangulating you saw in Chapter Four—dividing and conquering to the limit.* At the same time, it extends the notion of what can have area—but only so far. Too many holes, too close together, and area evaporates. That would be true, for instance, of the famous Sierpinski Gasket, made by deleting the 'midpoints triangle' from an original triangle, then the midpoints triangles from those remaining—and so on, forever. The area of what's left turns out to be zero (a quarter of what was

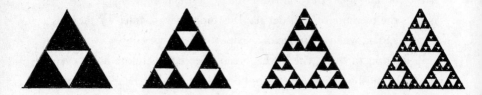

there at each stage is removed at the next). Yet since $0+0=0$, you could, in a frolicsome mood, extend the Pythagorean Theorem to a right triangle sporting Sierpinski Gaskets on its sides. Or has our drama just turned absurdist?

---

* A proof by integration that the areas of similar figures are to one another as the squares of their corresponding linear measures is in the appendix to this chapter.

TRIANGLES OTHER THAN RIGHT

Dog owners grow to resemble their dogs, and mathematicians the abstract pets they care for. They even lose their specific names and become generic, as did Pappus. What we have of him is that surname, chunks of his *Synagoge (The Collection)*, and the date of an eclipse he mentions: A.D. 320. He taught in Alexandria, and broadened the Pythagorean Theorem to *all* triangles by proving that if parallelograms were built on two sides, there would be a parallelogram on the third side whose area was the sum of theirs.

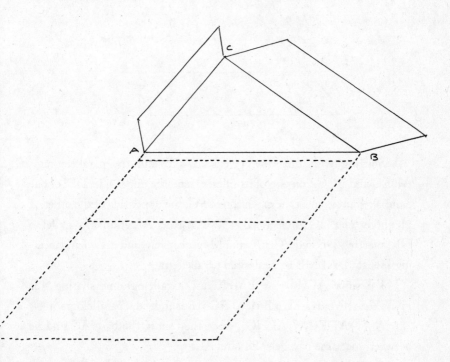

Here is how an ironically modern turn of thought might show this. The area of those two given parallelograms is some positive number, call it A. On the third side let a parallelogram grow steadily larger, from an area of zero (when, collapsed, its long sides coincide) to a size with area well past A. Somewhere along the way will have been the parallelogram you sought, with area exactly A.

You should feel as cheated by this as you would were you to ask a stranger whether he could give you directions to a renowned restaurant, and he answered that indeed he could, and walked away. "I meant . . . ," you shout after him, in vain. That's the last time you'll ever visit *this* city.

Pappus had no irony in his soul. These are his directions for building exactly the renowned parallelogram you hoped for. Everything turns on transitivity among parallelograms.

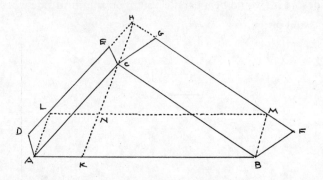

Starting with the same givens, ΔACB and the parallelograms ADEC and BCGF on two of its sides, extend the upper side, DE, of one parallelogram to meet the extension of FG, the upper side of the other, at H, and extend HC to meet AB at K. Construct AL and BM parallel to HK, meeting DE and FG at L and M respectively, and draw LM, meeting HK at N. ALMB is the desired parallelogram.

For since ALHC and BMHC are parallelograms, sharing side HC, AL ≅ BM and AL ∥ BM, so ALMB is indeed a parallelogram.

Now |ADEC|=|ALHC|, since they share the base AC and lie between the same parallels, DE and AC.

But |ALHC|=|ALNK|, since they share the base AL and lie between the same parallels, AL and KH.

By transitivity, |ADEC|=|ALNK|, and one of our original parallelograms has the same area as part of the desired one.

In the same way, |BFGC|=|BMNK|: the second of the original parallelograms has the same area as the remainder of the desired one, and we're done.

Pappus's Theorem not only generalizes the Pythagorean but—like Ptolemy's—generates it, since it is proven independently (and the Pythagorean follows when C is taken as a right angle and the parallelograms as squares). It is its own grandpa. Those nested cones of generalization we spoke of—how long the view down them in either direction, reminding us that in math, as in most things, we always stand midmost. Unlike our simple model of causality, however, these endless cones needn't lie in a straight line: their array might, as here, cavort in self-intersections.

△ △ △

We mentioned that the Pythagorean Theorem startles us in part because we estimate area so badly: the square on the hypotenuse still doesn't look big enough to contain the squares on the two sides. What would you make of someone who saw that in *any* triangle, the sum of the squares on the two sides added up to *twice* the square on *half* of the third, plus *twice* the square on the median to that side? "Prodigy" comes to mind, with its hint of the uncanny. Apollonius of Perga, then, was such a prodigy, for this, they say, is his theorem. He moved among ratios and lines in a triangle as casually as we walk around the neighborhood. His was Perga, in Pamphylia (on the southwestern coast of Turkey), and later, Alexandria; and the time—about 240 B.C.—puts him in the generation after Archimedes.

For the sake of ease we'll give his proof algebraically, although this attenuates the magic of seeing it through the play of shapes.

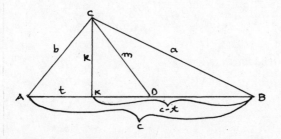

In △ABC, with sides a, b, and c as shown, drop the median m meeting AB at D, and the altitude k from C to AB, meeting it at K. Let AK = t, so that KB = $c - t$.

Apollonius wishes to prove that $a^2 + b^2 = \dfrac{c^2}{2} + 2m^2$.
By the Pythagorean Theorem,

$$a^2 = k^2 + (c - t)^2 = k^2 + c^2 - 2ct + t^2,$$

and

$$b^2 = k^2 + t^2,$$

so

$$a^2 + b^2 = 2k^2 + 2t^2 - 2ct + c^2$$

which (splitting c^2 in half and arranging the terms conveniently) turns into

$$a^2 + b^2 = \frac{c^2}{2} + 2k^2 + 2t^2 - 2ct + \frac{c^2}{2}.$$

Now again by the Pythagorean Theorem,

$$m^2 = k^2 + \left(\frac{c}{2} - t\right)^2 = k^2 + \frac{c^2}{4} - ct + t^2,$$

hence

$$2m^2 = 2k^2 + \frac{c^2}{2} - 2ct + 2t^2,$$

and we indeed have

$$a^2 + b^2 = 2m^2 + \frac{c^2}{2}.$$

Talk about generalization's nested cones—this theorem has a yet broader form, with D roaming as it will over AB. Matthew Stewart in Rothesay, off the distant Scottish coast, some time in the 1730s came up with a theorem for this more general setting that now proudly flies his name[1]—Stewart's Theorem: If D is any point on c, dividing it into the segments m and n, and d is the line from C to D, then

$$b^2 m + a^2 n = c(d^2 + mn).$$

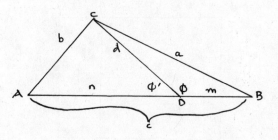

Stewart's proof (you'll find it in this chapter's appendix) follows from applying to $\triangle ADC$ and $\triangle CDB$ the most famous of all extensions of the Pythagorean Theorem to any triangle: The Law of Cosines.

You may think Apollonius's Theorem frivolous and Stewart's extension of it an effete variation. Yet neither is an idle étude—each is a significant sharpening of the mathematical outlook: a triangle's three sides and the various lines that go with them (altitudes, medians, and so on) depend on a few defining relations, one of which Apollonius gives for medians, and Stewart for the more general 'Cevian' (any line from the vertex to a triangle's opposite side): a relation fundamental enough to determine all the rest for the triangle you see and the whole class of triangles congruent to it. How freedom lines up with necessity: the sort of acorn you carry off and hope to live on.

△ △ △

This scary-sounding Law of Cosines, along with Driver's Ed, is what definitively tells the recent middle school graduate that the world of higher learning has now begun. We know it in terms of trigonometry—the art of the ratio, developed by Arabic mathematicians:

$$c^2 = a^2 + b^2 - 2ab \cdot \cos \gamma,$$

where γ is the angle between sides a and b.

A millennium earlier, however, Euclid proved this with stunning simplicity. What's more, his proof resurrects for us the Babylonian Box. You'll remember we said in Chapter Two (p. 15) that the box, rightly thought of, showed that $(a+b)^2 = a^2 + 2ab + b^2$.

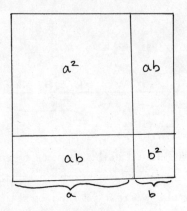

Showed—but didn't prove, which Euclid, being a Greek, required. His proof (II.4 in his *Elements*) just adds a diagonal, making slews of equal

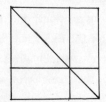

angles where it crosses pairs of parallel lines, letting him deduce that what look like squares are, and that the remaining oblongs are indeed congruent rectangles.*

With this result clipped to his belt and the Pythagorean Theorem in his pocket, Euclid now sets to work on the Law of Cosines. Here—in current terminology—is his elegant proof in the obtuse case (he follows it with a similar proof for the acute).

---

* Then as now the spirit of the diagonal hovered helpfully over our mathematical affairs.

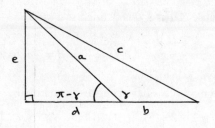

EUCLID (II.12): $c^2 = a^2 + b^2 + 2bd$

PROOF:

By II.4, $(b+d)^2 = b^2 + d^2 + 2bd$.

Add e^2 to both sides, and rearrange the right:

$$(b+d)^2 + e^2 = d^2 + e^2 + b^2 + 2bd$$

By the PT, $\qquad\qquad \downarrow \qquad \downarrow$

$$c^2 \quad = \quad a^2 + b^2 + 2bd.$$

To translate this into our familiar "law",* recall from your days of trigonometry that

$$\cos(r - s) = \cos r \cdot \cos s + \sin r \cdot \sin s,$$

so (knowing that $180° = \pi$ radians, and that $\cos \pi = -1$ and $\sin \pi = 0$),

$$\cos(\pi - \gamma) = \cos \pi \cdot \cos \gamma + \sin \pi \cdot \sin \gamma = -\cos \gamma,$$

---

* The shift from working with areas of rectangles to functions of angles is part of the almost glacial movement we have detected, time and again, from a static to a dynamic view of mathematics. While it is true that Euclid speaks in these theorems of choosing "a random point" (such as D), the figure, as it were, then follows this choice and so remains at rest, while looking at it in terms of angles and ratios animates it. As ratios came to take on the status of numbers, Being and Becoming—the figure and ground of content and context—definitively exchanged places.

hence (because $\cos(\pi - \gamma)$ is defined as a ratio),

$$-\cos\gamma = \cos(\pi - \gamma) = \frac{d}{a} \Rightarrow d = -a\cos\gamma,$$

so (multiplying both sides of this last equation by $2b$),

$$2bd = -2ab\cos\gamma,$$

giving us

$$c^2 = a^2 + b^2 - 2ab \cdot \cos\gamma.$$

One consequence of this result is the converse of the Pythagorean Theorem: if $a^2 + b^2 = c^2$, the triangle whose sides these are is right. For if not, the angle in question being acute or obtuse would add a non-zero term to this sum (it needs of course no spirit from the trigonometric deep to tell us this: Euclid easily proves (I.48) the converse of the Pythagorean Theorem from the Theorem itself). Nothing happens in isolation: we will soon need this converse.

△ △ △

You've just seen from the law of cosines not only *that*, but *why*, c^2 is greater than $a^2 + b^2$ in an obtuse, less than it in an acute, and equal to it in a right triangle. With the change in viewpoint from the static to the dynamic, it's tempting to hook this trio up to the paired inequalities of the angles: $\gamma > \alpha + \beta$, $\gamma < \alpha + \beta$, and $\gamma = \alpha + \beta$ respectively.

Edsger Dijkstra, among the last century's leading pioneers in computer science, found a more succinct way of putting this (using '⇔' to mean 'if and only if'):

$$c^2 > a^2 + b^2 \quad \Leftrightarrow \quad \gamma > \alpha + \beta$$
$$c^2 = a^2 + b^2 \quad \Leftrightarrow \quad \gamma = \alpha + \beta$$
$$c^2 < a^2 + b^2 \quad \Leftrightarrow \quad \gamma < \alpha + \beta.$$

Looking for greater concision, he rewrote:

$$a^2+b^2-c^2<0 \quad \Leftrightarrow \quad \alpha+\beta-\gamma<0$$
$$a^2+b^2-c^2=0 \quad \Leftrightarrow \quad \alpha+\beta-\gamma=0$$
$$a^2+b^2-c^2>0 \quad \Leftrightarrow \quad \alpha+\beta-\gamma>0$$

For now he could rethink these expressions in terms of their *signs*—negative, zero, or positive—and (in the grip of this frenzy for condensing) abbreviate even 'sign' to the sign 'sgn', and so conclude:

$$\mathrm{sgn}\,(a^2+b^2-c^2) \Leftrightarrow \mathrm{sgn}\,(\alpha+\beta-\gamma).$$

There's our whole book so far, freeze-dried to half a line of squiggles.

Dijkstra wasn't content with a mere summary. In fact he seems not to have been content with most things, making him an enfant terrible to many, an awful child to more. One of his colleagues spoke of him as a

giant, adding: "But intellectual giants are odd. They distort the usual laws of perspective: as you get closer to them, they seem smaller." His oddness included indulging in a running fantasy about a company called Mathematics Inc., of which he was the chairman. It manufactured theorems, keeping their proofs a trade secret (it had also produced The

Standard Proof of the Pythagorean Theorem, which outsold all the incompatible existing proofs).

In a memorial for him, another colleague wrote: "Opinions, different from his, met with his greatest disapproval, and he related to them in a famously obnoxious manner." Arrogance in computer science is apparently now measured in nano-Dijkstras. Yet while he cultivated rudeness toward the world, his small band of followers thought him compassionate, a man who would drop in unannounced of an evening to chat for an hour or two—a grammarian in sandals.

In any case, Dijkstra now set out in reality to prove his Standardized Statement of the Pythagorean Theorem. He drew a picture of an obtuse triangle ABC with its obtuse angle, γ, at C:

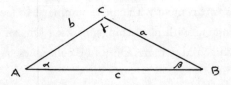

and since $\gamma > \alpha + \beta$, $\alpha + \beta - \gamma < 0$, so its sign, $\mathrm{sgn}(\alpha + \beta - \gamma)$ is negative.

He then created two triangles, 1 and 3, similar to $\triangle ABC$ and inside it, as shown, with an isosceles triangle 2 between them:

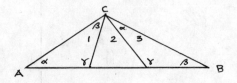

He no sooner did this than he regretted it: "No cheers at all for that stage of the argument which forced us to resort to a picture. Pictures are almost unavoidably overspecific." But he bravely went on: since $\Delta 1$ and $\Delta 3$ don't cover $\triangle ABC$,

$$|\Delta 1| + |\Delta 3| < |\triangle ABC|, \text{ i.e., } |\Delta 1| + |\Delta 3| - |\triangle ABC| < 0,$$

so its sign too is negative, hence

$$\mathrm{sgn}\,(|\,\Delta 1\,|+|\,\Delta 3\,|-|\,\Delta \mathrm{ABC}\,|)=\mathrm{sgn}\,(\alpha+\beta-\gamma).$$

Dijkstra is almost there. Since $\Delta 1 \sim \Delta 3 \sim \Delta \mathrm{ABC}$, ratios of their areas to the squares of their sides are equal:

$$\frac{|\,\Delta_1\,|}{b^2}=\frac{|\,\Delta_3\,|}{a^2}=\frac{|\,\Delta \mathrm{ABC}\,|}{c^2},$$

so

$$\mathrm{sgn}\,(a^2+b^2-c^2)=\mathrm{sgn}\,(|\,\Delta 1\,|+|\,\Delta 3\,|-|\,\Delta \mathrm{ABC}\,|)=\mathrm{sgn}\,(\alpha+\beta-\gamma),$$

as desired—in the obtuse case, at least. Dijkstra notes that were $\alpha+\beta=\gamma$, $\Delta 2$ would disappear, so that $|\,\Delta 1\,|+|\,\Delta 3\,|=|\,\Delta \mathrm{ABC}\,|$; and were $\alpha+\beta>\gamma$, $\Delta 1$ and $\Delta 3$ would overlap—a situation he carefully avoids drawing, since, he says, there would be nine possible cases—but in each, $|\,\Delta 1\,|+|\,\Delta 3\,|>$ $|\,\Delta \mathrm{ABC}\,|$.

Well, what do you make of all this? Is it just another, rather awkward, proof of the law of cosines, or something much more? Dijkstra wrote: "We have proved a theorem, say, 4 times as rich as [the usual formulation of the Pythagorean Theorem]." What exactly is this richness? Does it lie in having reduced the statement to a bit of code? Elsewhere Dijkstra said that "the traditional mathematician recognizes and appreciates mathematical elegance when he sees it. I propose to go one step further, and to consider elegance an essential ingredient of mathematics: if it's clumsy, it's not mathematics."

Is this mathematics? We come back to our discussion of elegance in Chapter Five, where (to put it in our present terms) we suggested that elegance and economy may have the same sign, but aren't necessarily equal. Even if they were, all those omitted cases clutter up the background of this proof, and a mere word count, or inventory of its steps, shows it less economical than Euclid's. But something much more important, we think, is involved. For all the statement's concision, it has been torn from the history, associations, intuitions, and implications of the Pythagorean Theorem, and made into a cell in cyberspace that discloses nothing. It is as if you wanted to hear a quartet movement of Mozart's and were played instead the chord from which its harmonies and melodies unfolded. Neither math nor music is an excursion in syntax, but in semantics.

We'll leave the last words to Dijkstra himself. He ends his proof with this sardonic, if not quite English, reflection: "I am in a paradoxical situation. I am convinced that of the people knowing the theorem of Pythagoras, almost no one can read the above without being surprised at least once. Furthermore, I think that all those surprises relevant (because telling about their education in reasoning). Yet I don't know of a single respectable journal in which I could flog this dead horse." Well, there was always Mathematics Inc.[2]

A PLANE OTHER THAN THE REAL

So precariously is mathematics balanced on the edge of tautology that the gentlest push can tumble it into the depths. Renaming is such a push (since it really stands for looking at one thing from two points of view), as when we call our triangle's sides no longer a, b, and c but $\sin \theta$, $\cos \theta$, and 1,

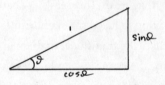

so that

$$\sin^2 \theta + \cos^2 \theta = 1.$$

In the best tales, the door to the secret garden blends into the bricks of its wall, and tautologies are the bricks of mathematics. Look at the garden the miraculous Swiss mathematician Leonhard Euler fell into when he opened this door.[3] He first factored the new equation into

$$(\cos\theta + i\sin\theta)(\cos\theta - i\sin\theta) = 1$$

—a jaw-dropping move—from which he swiftly developed the infinite series you saw in Chapter Five:

$$\sin x = x - \frac{x^3}{3!} + \frac{x^5}{5!} - \frac{x^7}{7!} + \cdots$$

$$\cos x = 1 - \frac{x^2}{2!} + \frac{x^4}{4!} - \frac{x^6}{6!} + \cdots$$

But there were orchids yet more fantastic growing here, which he and others cultivated, such as his astonishing formula,

$$e^{i\theta} = \cos\theta + i\sin\theta.$$

The garden we have found ourselves in is the complex plane, where so much that was unsettled on the real is completed, and symmetry restored. An nth degree polynomial, for example, has at most n *real* roots—but all n of them spread through the *complex* plane—and in patterns that would delight a Capability Brown: 1 has two real square roots, 1 and –1; three complex cube roots $\left(1, \dfrac{-1 + i\sqrt{3}}{2}, \dfrac{-1 - i\sqrt{3}}{2}\right)$; four complex fourth roots; and so on: exactly n complex nth roots, arranged as the vertices of a regular n-gon:

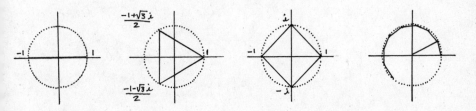

These all are the offspring of $c = \sqrt{a^2 + b^2}$ and $i = \sqrt{-1}$. We will wander in this garden later, to collect some of its most exotic species: flowers whose corollas tunnel to another world.

DIMENSIONS OTHER THAN TWO

Because the Pythagorean Theorem concerns areas, we think of one plane or another as its natural home. But inquiry is restless and imagination boundless, so very soon you ask: what about the 'body diagonal' of a rectangular box?

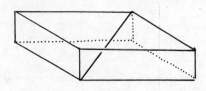

Draw in the hypotenuse e on the floor of this box. Then on the plane

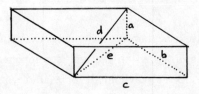

made by a, e, and d,

$$d^2 = a^2 + e^2.$$

But back on the floor,

$$e^2 = b^2 + c^2,$$

so

$$d^2 = a^2 + b^2 + c^2.$$

△ △ △

Was ever so wide an insight leapt with so little effort? Perhaps the unused energy you stored up for it has set off a wild surmise in your parietal cortex (where we do our counting): a sum of two squares in two dimensions, of three in three . . . But let your eye restrain your brain for a moment by looking back to seventeenth-century Ulm, where the Master Reckoner Johannes Faulhaber not only discovered a different spatial generalization of the Pythagorean Theorem, but saw how it would

lead him to a deeper understanding of 666, which he already knew was divine rather than diabolic. While the margins of our minds are too narrow to contain Faulhaber's kabbalistic proof, we will unveil to you his rectilinear pyramid.

"Analogy", said his contemporary Johannes Kepler, "is my most faithful teacher." Moving by analogy from squares on the *lengths* of a right triangle's *sides* to squares on the *areas* of an orthonormal pyramid's *faces*, he came up with this beautiful insight: given such a pyramid (three of its edges meeting at right angles, as in the corner of a box), the sum of the squared areas of these faces (call them A, B, and C)

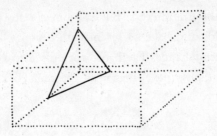

would equal the square of the area of the last, 'hypotenuse', face, D (i.e., ΔXWY): that is,

$$|A|^2+|B|^2+|C|^2=|D|^2.$$

If not quite up there with 666, still pretty wonderful. Here's a more precise statement of the theorem, and its proof.

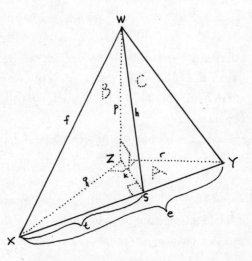

THEOREM: In a tetrahedron, three of whose edges (p, q, r) meet at right angles at vertex Z, if the areas of the three faces meeting at Z are $|A|$, $|B|$, and $|C|$, then with $|D|$ the area of the fourth ('hypotenuse') face,

$$|A|^2+|B|^2+|C|^2=|D|^2.$$

PROOF:

1. With the diagram as lettered—and D the front, 'hypotenuse', face, WXY—from Z construct $k \perp e$ at S, making segment $XS = t$. Construct WS and label it h.
2. Notice that $q^2 + r^2 = e^2$.
3. Now $k^2 + t^2 = q^2$, so $k^2 = q^2 - t^2$.
4. $p^2 + q^2 = f^2 \Rightarrow p^2 = f^2 - q^2$.
5. Hence $k^2 + p^2 = f^2 - t^2$.
6. But $\angle SZW$ is a right angle, so $k^2 + p^2 = h^2$.
7. Hence (substituting from step 5), $h^2 = f^2 - t^2 \Rightarrow$ $h^2 + t^2 = f^2 \Rightarrow h \perp t$.

$$\downarrow$$

converse of the Pythagorean Theorem

8. So $|D| = \dfrac{eh}{2} \Rightarrow 4\,|D|^2 = e^2 h^2 = (q^2 + r^2)(p^2 + k^2)$

$$= q^2 p^2 + q^2 k^2 + r^2 p^2 + r^2 k^2$$

$$\Rightarrow 4|D|^2 = q^2 p^2 + r^2 p^2 + k^2(q^2 + r^2)$$

$$= q^2 p^2 + r^2 p^2 + k^2 e^2$$

$$\downarrow \qquad \downarrow \qquad \downarrow$$

$$4|B|^2 + 4|C|^2 + 4|A|^2$$

$$\Rightarrow |D|^2 = |A|^2 + |B|^2 + |C|^2.$$

Good ideas are born over and over because, like Pythagoras, they would show their immortality through reincarnation. Descartes may have co-fathered Faulhaber's discovery; then in late eighteenth-century France its new parents were d'Amondans Charles de Tinseau and—independently, it seems—Jean-Paul de Gua de Malves: the first editor of the great *Encyclopédie*, and the man whose name is usually attached to this theorem (The Return of Stigler's Law). But the best ideas have progeny, not just doubles, and now we can let loose that numerical impulse so recently restrained.

We said rather cryptically that Descartes might have shared with Faulhaber the parentage of this theorem. In the winter of 1619 the two may well have met in Ulm. Local legend has it that one night, dazzled

by his brilliance, Faulhaber reached out and touched his guest to make sure that he was human and not an angel. Wouldn't you have done as much, were your visitor to have mentioned that this theorem also held in four dimensions? Descartes wrote, in his *Private Thoughts* (*Cogitationes privatae*), dating from 1619–21:

> This demonstration comes from the Pythagoreans and can also be extended to quantities of four dimensions. There the square of the solid opposed to the right angle is the squares from the other four solids altogether. To this let there be the example of progression in numbers 1, 2, 3, 4; in right angles of two, three, four lines.[4]

Yet what does "let there be the example" mean? Might faithful analogy not have been sent to damn us? The restraint we had exercised before came from rightly recognizing that fantasy can lead imagination astray; the informing particularity of things can vanish at a touch of abstraction. The architectural instinct carries us too easily past detail, as *would* makes *could*, *could* makes *might*, and *might* makes *must* grow hazy. Hilbert's more formalist followers, and people like Dijkstra, help in making syntax abet semantics, so that *how we say* will enhance *what we think* (though then they err by narrowing correction to discipline—as if they had emptied the perspective grids in a Vermeer of their lively content, leaving behind mere Mondrian).

We can go past Descartes and generalize this theorem to five—to six—to spaces of any dimension you like. We can deduce that the hypotenuse 'face' of such a 'pyramid' in, say, a thousand and one dimensions has an 'area' equal to the sum of that of the other 'faces' (where we need all those raised eyebrows of 'so-called' because the faces, for instance, are—as in two and three dimensions—one dimension down from that of the space the object lives in: so here, thousand-dimensional faces with their multi-dimensional equivalent of area). But we'll first have to understand how to think about (rather than just be stunned by) all this, since seeing it is out of the question. We'll have to insure that we have been entertaining an angel, not a devil, unawares.

The insurance comes from that cautious run-up called induction, which, like the high jumper's measured steps, will hurl us over the bar (how it works is explained in the appendix to this chapter). It was born and reborn from at least the eleventh through the seventeenth centuries, but only in 1912 did the great French mathematician Henri Poincaré see how it would let us precisely grasp higher dimensions. The question with induction is always: have you a way of going uniformly from any stage to the next? Poincaré made use of a very simple observation: two points on a line will be separated by a cut at a point between them. Since the line is one-dimensional and a point has dimension zero, perhaps the line's dimension depended on the dimension of what was needed to cause the separation: 1 was the successor of 0.

Did this work for a two-dimensional plane? Given any two points on it, they could be separated from one another only by cutting a closed curve (which has dimension one) around one of them—and 2 is the successor of 1. Likewise in three dimensions, it would take a two-dimensional surface (like the skin of a sphere) to separate any two points—no lower-dimensional boundary would always suffice—and 3 is the successor of 2.

Rather than a *proof* by induction, therefore, Poincaré made this *definition* by induction: a space is n-dimensional if any two points in it can be separated by an $n-1$ dimensional subset of it, and no lower-dimensional subset would always work.

What guarantees that this inductive definition is right? We're past the child's fiat "Anything I say three times is true." As in all matters of religion, what's left is justification by faith and by works. Our faith rests on uniformity: the picture Poincaré's definition builds up makes the next dimension arise from the prior in such a way that the tools and insights we had before will extend smoothly. If you say this is all very well, but the higher dimensions we thus get are just the playing fields for our game, as arbitrary in their plan as are the game's rules, we would agree—but add that this seems to be the only game in town.

Justification by works comes from seeing how this understanding of dimension enlightens and predicts in physics: one degree less separated from what we think of as the real world. We need three dimensions to locate a particle in this world—locating its distance in height, length, and breadth from a fixed center. These three axes we conveniently think of as at right angles to one another. This picture accords with Poincaré's, for he had three-dimensional Euclidean space in mind as his model, and then broadened it to achieve his topological viewpoint (which has evolved greatly since him). Once the particle moves, however, we need to keep track of the forces causing its motion—momenta in those three directions. Since these are independent of one another and of the directions, we'll need three more axes, each at right angles to all the rest, in order to pin down the moving particle—altogether the six dimensions of 'phase space'. You can imagine adding more criteria that you want to keep track of, and new dimensions, one at a time, for each. As in the moral realm, what was so recently unthinkable has become a matter of course.

We're prepared now for the modern generalization[5] of the Pythagorean Theorem to any dimension n. Let's first choose the right words for so overweening a project. What was a triangle in two and a tetrahedron in three dimensions is called a *polytope* in higher dimensions. We'll let P stand for this orthogonal polytope—i.e., one with that right-angled vertex (though you might prefer to think of P as standing for our pyramid, no matter what dimension it is at home in).

A triangle has legs, a pyramid faces—each therefore with its kind of boundary one dimension down from that of the figure itself. The $n-1$ dimensional equivalents of these for an n-dimensional P are called *facets*. What was a right triangle's *hypotenuse*, and the face opposite the pyramid's right-angled corner its *hypotenuse face*, will now accordingly be the *hypotenuse facet*. And while we worked with a facet's *area* in two dimensions and *volume* in three, the general term in any dimension is *content*.

So the stunning generalization we hope to prove is that for P in any dimension n, the square of the hypotenuse facet's content equals the sum of the squares of the contents of its other facets. This hope will become a reality in the chapter's appendix, should you choose to stretch your legs—or really your mind—there. For while the ideas, tactics, and strategies are thoroughly human, the language that bears them has a concision that takes some getting used to, and abbreviations that are immensely helpful in the end, but intimidating when you begin. And if you choose *not* to go for that exhilarating run, your hope may reliably be replaced by faith.

FORMS OTHER THAN SHAPES

Pythagoras's soul moved from body to body, the Pythagorean Theorem from plain to ever more rarefied triangular embodiments. The natural movement now is for shape to distill wholly away to the single malt of number: the ultimate generalization whose intoxicating fumes layer the mind's upper air. But this will need a chapter of its own.

APPENDIX

(A) We promised a proof that if T_1 and T_2 are similar closed curves on straight line segments, then their areas are to one another as the squares of those segments.

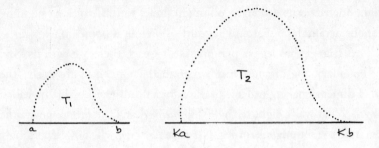

PROOF:

We want to show that $\int_{ka}^{kb} kf\left(\dfrac{x}{k}\right)dx = k^2\int_a^b f(x)dx$,

i.e., that $\int_{ka}^{kb} f\left(\dfrac{x}{k}\right)dx = k\int_a^b f(x)dx$.

(Note: why $f\left(\dfrac{x}{k}\right)$? Because in scaling up the whole by k, we have to scale down the inputs to f by $1/k$ to obtain a similar graph.)

Letting $g(x) = \dfrac{x}{k}$, $g'(x) = \dfrac{1}{k}$, and we have

$$\int_{ka}^{kb} f\left(\frac{x}{k}\right)dx = \int_{ka}^{kb} kf\left(\frac{x}{k}\right)\left(\frac{1}{x}\right)dx = \int_{ka}^{kb} kf(g(x))g'(x)dx$$

$$= k\int_{ka}^{kb} f(g(x))g'(x)dx = k\int_a^b f(u)du = k\int_a^b f(x)dx.$$

$$\uparrow \qquad\qquad \uparrow$$

integration   change of variables
by substitution

(B) This is Matthew Stewart's proof of his theorem, which you saw in the chapter: $b^2m + a^2n = c(d^2 + mn)$.

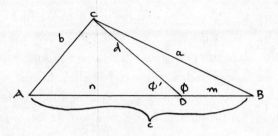

1. Call the angle between d and m φ, and its supplement, between d and n, φ'.

2. Because φ' is the supplement of φ, trigonometry tells us that $\cos \varphi' = -\cos \varphi$.

3. By the law of cosines,
   $$a^2 = m^2 + d^2 - 2dm \cos \varphi.$$

4. Again by the law of cosines,
   $$b^2 = n^2 + d^2 - 2nd \cos \varphi' = n^2 + d^2 + 2nd \cos \varphi \text{ (by step 2)}.$$

5. Multiplying step 3 by n:
   $$na^2 = nm^2 + nd^2 - 2dmn \cos \varphi.$$

6. Multiplying step 4 by m:
   $$mb^2 = mn^2 + md^2 + 2dmn \cos \varphi.$$

7. Adding steps 5 and 6:
   $$na^2 + mb^2 = nm^2 + mn^2 + nd^2 + md^2 - 2dmn \cos \varphi + 2dmn \cos \varphi$$
   $$= nm^2 + mn^2 + (m+n)d^2$$
   $$= nm(m+n) + (m+n)d^2$$
   $$= (m+n)(nm+d^2)$$
   $$= c(nm+d^2).$$
   $$\text{Q. E. D.}$$

(C) Here is your kit for induction. If dominoes are set up vertically in a line on a table and you push over the first, all the rest will fall down in sequence—*as long as you've set them up so that each is less than a domino-length from the next.* Think of the natural numbers (1, 2, 3, . . .) as these dominoes, and suppose you want to prove that some statement is true for each and every one of them. You first prove it is true for the first number, 1—that's like pushing over the first domino. Then (and this is the great idea), prove that *if your statement is true for any one of the numbers, then it will be true for the next.* That's like making sure that the dominoes are close enough together to fall when the first is tipped. If you manage both steps (proving your statement for $n = 1$, then proving that *if* it holds for the number k, it must also hold for $k + 1$), you will have proved it true for all natural numbers: a proof by induction.

Suppose, for instance, you want to prove that the sum of the internal angles of an n-sided polygon is $(n - 2)180°$.

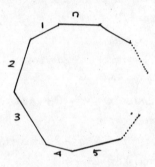

Well, this is certainly true for a triangle (where $n = 3$)—though this needs to be established ahead of time, and deductively. Now assume this statement is true for a k-sided polygon: i.e., assume that its interior angles add up to $(k - 2)180°$. *Using this assumption*, prove that the statement is also true for a $(k + 1)$-sided polygon: i.e., that its interior angles add up to $(k + 1 - 2)\,180°$, or $(k - 1)180°$.

You may object at this point that we're assuming what we're setting out to prove! Not really: We're *assuming* only that it holds for a single number, k—and then somehow or other we'll *prove* that it holds also just for the next, $k + 1$.

So here's the picture of our assumption: a k-sided polygon, with its angles adding up to $(k - 2)\,180°$.

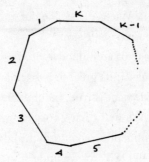

Let's draw a $(k+1)$-sided polygon:

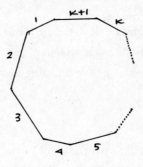

and (here's the cleverness) nest a k-sided one within it:

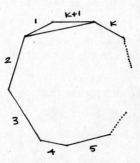

Behold! (as the second Bhaskara would say): by our "inductive assumption", the angles of the k-sided polygon add up to $(k-2)\,180°$; and perched on top of it, a solitary triangle, whose angle sum is $180°$. Altogether, then, the angle sum of this composite figure is $(k-2)180° + 180° = (k-1)180°$. Now remove the interior line that separated this triangle from the shape below it. We lose one side but gain two, so have indeed $k+1$ sides, and the required angle sum.

The great advantage of an inductive proof is the relative ease with which it can often be made; its great drawback is that the proof substitutes *how* for *why* a statement is true, indubitably establishing *that* it is, by the domino effect. We never have nor ever will see a trillion-sided polygon, yet are as confident as about anything in life that its interior angles add up to $(10^{12} - 2)\,180°$, for we understand the gearing. So, after the initial astonishment at the figures emerging on the hour from Munich's Glockenspiel, there may be appreciation but *there will be no surprises.*

This peculiar situation reflects the fundamentally structural nature of mathematics and its objects: abstraction, generality, and truth ripple through it, and them, because their meaning isn't (some would argue) to be sought elsewhere.

(D) We end this appendix with the proof of the Pythagorean Theorem in n dimensions—a generalization of Faulhaber's. A word first about those notational conventions used to simplify the algebra.

Exponents describe multiplication (of a number by itself), but seem to behave additively. $a^2 \cdot a^3 = (a \cdot a)(a \cdot a \cdot a) = a^5$, and in general, $a^r \cdot a^s = a^{r+s}$.

This has the helpful consequence that $\dfrac{a^5}{a^3} = \dfrac{a \cdot a \cdot a \cdot a \cdot a}{a \cdot a \cdot a} = a^2$ suggests $\dfrac{a^r}{a^s} = a^{r-s}$. This in turn means that $\dfrac{a^3}{a^5} = \dfrac{a \cdot a \cdot a}{a \cdot a \cdot a \cdot a \cdot a} = \dfrac{1}{a^2} = a^{-2}$, defining negative exponents.

In particular, $\dfrac{a}{a} = 1$ gives us $a^{1-1} = a^0 = 1$.

This notation also forces the meaning of fractional exponents on us. If $a^x \cdot a^x = a^1$, we must have $2x = 1$, so $x = \dfrac{1}{2}$. But when a number times itself is a, that number must be $\sqrt{a}$. Hence $a^{\frac{1}{2}}$ means $\sqrt{a}$, and in general, $a^{\frac{1}{n}} = \sqrt[n]{a}$.

Combining this with our first result, $a^{\frac{r}{s}} = \sqrt[s]{a^r}$.

The way we write exponents is a fine example of notation and meaning leading each other on.

The other notation we use here is the abbreviation Σ (sigma, the upper-case Greek S) for 'sum'. Instead of writing $a_1 + a_2$ we could write: "the sum of a_j, as the index j goes from 1 to 2"—abbreviated

$$\sum_{j=1}^{2} a_j$$

This seems an absurd convention until you get to longer—and arbitrarily long—sums:

$$\sum_{j=1}^{7} a_j$$

is briefer and more transparent than $a_1 + a_2 + a_3 + a_4 + a_5 + a_6 + a_7$, and certainly

$$\sum_{j=1}^{n} a_j$$

is easier on hand, eye, and brain than $a_1 + a_2 + a_3 + \cdots + a_{n-1} + a_n$.

If something is to be done to each term before adding (such as raising each to the power −2), we can bring this into the notation:

$$\sum_{j=1}^{n} a_j^{-2}.$$

Now we're ready for the proof itself.

A triangle in 2-space has 3 sides, a tetrahedron in 3-space has 4 faces, and P in n-space has $n+1$ facets, one of which is the hypotenuse facet F_h. So what we want to prove is that

$$|F_h|^2 = |F_1|^2 + |F_2|^2 + \cdots + |F_n|^2,$$

or, using the shorthand explained above,

$$|F_h|^2 = \sum_{j=1}^{n} |F_j|^2$$

You know that the volume (area) of a triangle in 2-space is half the base times the height, as here:

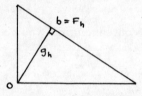

Looked at more generally, this base is the hypotenuse face, F_h, so

$$V_2 = \frac{1}{2}|F_h| \cdot g_h$$
$$\downarrow$$

length of the altitude
to F_h from O.

Of course we could have gotten the same volume by taking half the content (i.e., length) of another side times the altitude to it:

$$V_2 = \frac{1}{2}|F_1| \cdot g_1 \qquad = \qquad \frac{1}{2}|F_2| \cdot g_2$$
$$\downarrow \quad \downarrow \qquad\qquad\qquad \downarrow \quad \downarrow$$

side a_1 side a_2 side a_2 side a_1

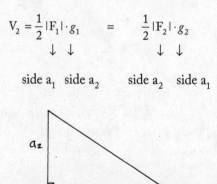

In $n=3$,

$$V_3 = \frac{1}{3}|F_h| \cdot g_h = \frac{1}{3}|F_1| \cdot g_1 = \frac{1}{3}|F_2| \cdot g_2 = \frac{1}{3}|F_3| \cdot g_3.$$

An inductive process generalizes both of these to

$$V_n = \frac{1}{n}|F_h| \cdot g_h = \frac{1}{n}|F_j| \cdot g_j \qquad (*)$$

Notice that we've used the subscripts to link a facet to the vertex it doesn't contain, and from which therefore its altitude is dropped.

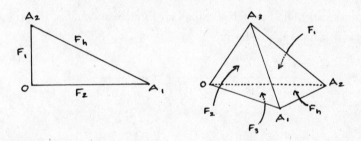

And if you wonder about that (*), it isn't only to decorate a meritorious achievement, but to give us a way to refer back to these equalities.

Our strategy will be to say the same thing, P's content, in two different ways, and reach our goal by sliding them together. Algebraic tactics along the way will be for the sake of this strategy.

We begin by going back to $n=2$ and recalling that $\triangle OSA_2 \sim \triangle A_1OA_2$,

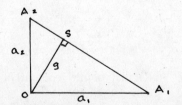

so $\dfrac{g}{a_2} = \dfrac{a_1}{A_1 A_2}$. But $A_1 A_2 = \sqrt{a_1^2 + a_2^2}$. Let's rewrite this in the more streamlined exponential notation:

$$A_1 A_2 = (a_1^2 + a_2^2)^{\frac{1}{2}}; \text{ so } \frac{g}{a_2} = \frac{a_1}{(a_1^2 + a_2^2)^{\frac{1}{2}}}.$$

This means that

$$g = \frac{a_1 a_2}{(a_1^2 + a_2^2)^{\frac{1}{2}}}.$$

We've expressed g in terms of the lengths of the triangle's sides, which will turn out to be just what we need. First, though, we have to transform our formula through six small algebraic manoeuvres, in order to put it in a more usable form:

$$g = \frac{a_1 a_2}{(a_1^2 + a_2^2)^{\frac{1}{2}}} = \frac{1}{\dfrac{(a_1^2 + a_2^2)^{\frac{1}{2}}}{a_1 a_2}} = \frac{1}{\left(\dfrac{a_1^2 + a_2^2}{a_1^2 a_2^2}\right)^{\frac{1}{2}}}$$

$$= \frac{1}{\left(\dfrac{a_1^2}{a_1^2 a_2^2} + \dfrac{a_2^2}{a_1^2 a_2^2}\right)^{\frac{1}{2}}} = \frac{1}{\left(\dfrac{1}{a_2^2} + \dfrac{1}{a_1^2}\right)^{\frac{1}{2}}} = \frac{1}{(a_2^{-2} + a_1^{-2})^{\frac{1}{2}}}$$

$$= (a_2^{-2} + a_1^{-2})^{-\frac{1}{2}} \Rightarrow g = \left(\sum_{j=1}^{2} a_j^{-2}\right)^{-\frac{1}{2}}$$

Generalizing to n dimensions, the altitude g_h from O to F_h is therefore

$$g_h = \left(\sum_{j=1}^{n} a_j^{-2}\right)^{-\frac{1}{2}} \qquad (**)$$

Substituting (**) in (*), the 'volume' (content) of our n-dimensional P is

$$V_n = \frac{1}{n} |F_b| \cdot \left(\sum_{j=1}^n a_j^{-2} \right)^{-\frac{1}{2}}.$$

Since we're interested in $|F_b|^2$, let's square this:

$$V_n^2 = \frac{1}{n^2} |F_b|^2 \cdot \left(\sum_{j=1}^n a_j^{-2} \right)^{-1}.$$

And solving for $|F_b|^2$,

$$|F_b|^2 = V_n^2 n^2 \sum_{j=1}^n a_j^{-2} \qquad (***)$$

Well and good—but we want to write $|F_b|^2$ in terms of the content of the other n facets, not the side-lengths a_j. Let's do this, and watch the concluding miracle happen.

Any facet F_j of P, other than F_b, contains O but not A_j, so the altitude to it is a_j.

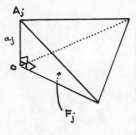

So that by (*),

$$V_n = \frac{1}{n} |F_j| a_j,$$

hence

$$V_n^2 = \frac{1}{n^2} |F_j|^2 a_j^2$$

Solving for a_j^2,

$$a_j^2 = \frac{V_n^2 n^2}{|F_j|^2},$$

so

$$a_j^{-2} = \frac{|F_j|^2}{V_n^2 n^2} \qquad (****)$$

Let's replace a_j^{-2} in (***) by this latest expression (****), first factoring the constant $\dfrac{1}{V_n^2 n^2}$ out of each term:

$$|F_b|^2 = \frac{V_n^2 n^2}{V_n^2 n^2} \cdot \sum_{j=1}^{n} |F_j|^2 = \sum_{j=1}^{n} |F_j|^2,$$

as desired!

CHAPTER SEVEN

Number Emerges from Shape

Amidst all the wreckage caused by $\sqrt{2}$, one Babylonian monument still towered up in the Pythagorean sky: those triples of whole numbers they had found which made sides of right triangles. At its base lay (3, 4, 5), with its multiples piled endlessly on. But you recall there were also (5, 12, 13), (7, 24, 25), and (11, 60, 61): in fact, the whole family of triples that came from fitting a pebbled gnomon to a pebbled square, so long as the number of pebbles in that gnomon was (despite its L-shape) itself a square number. So the gnomon around the square which has

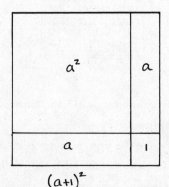

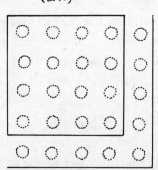

4 pebbles on a side has $9 = 3^2$ pebbles in it, and the gnomon fitted to a 24×24 square has $7^2 = 49$. As we would say, whenever a gnomon with $2a + 1$ pebbles is pushed up against a square with a^2 pebbles, the outcome $(a^2 + 2a + 1)$ is a square with $(a + 1)^2$ pebbles, and the ones that interest us are those where $2a + 1$ is a square number, though here arranged in an L (once again, it is mind that disembodies number).

We conjectured that the Old Babylonians had also discovered a second family of these triples, made by attaching *two* gnomons to a square, at its opposite corners, so long as the sum of their pebbles was also a perfect square—as when two gnomons with 32 pebbles each surround a square with $15^2 = 225$ pebbles, since $2 \times 32 = 64 = 8^2$, and $8^2 + 15^2 = 17^2$. Putting this too in our terms, $(a + 2)^2 = a^2 + 4a + 4$ (this $4a + 4$ is made up of the two gnomons each with $2a + 2$ pebbles).

It is awesome that our finite minds can encompass one, and then even two, such infinite families.

No tower but excites the urge to out-top it, as the Empire State Building pushed higher than the Chrysler, the World Trade Center and then the Sears Tower beyond both, and then the succession of Petronas Twin Towers, Taipei 101, Burj Khalifa—and Babel, past them all. Perhaps the ravages of the irrational elsewhere increased the need, among the Greeks, to find more infinite families of Pythagorean triples (as these came to be called): here among whole numbers, at least, rationality could work securely with its ratios. Or it may have been, if such study preceded Hippasus, that this made his revelation of the irrationality of $\sqrt{2}$ all the more devastating. But the skyscrapers of Manhattan are unmoved, though tunnels honeycomb the rock beneath them, and wonders are still to achieve. We now know that $\sqrt{2}$ isn't alone: there are incomparably many more irrationals than rationals. But this makes those whole number solutions blaze even more brightly among the dark numbers, as stars do in a universe made mostly of dark matter.

Yet how can we find *more* infinite families of Pythagorean triples, when there's nowhere else to attach gnomons to the inner square? *By turning away from shapes, for all their nursery warmth, and letting mind loose on number per se.* And when mind is let loose, its grasp tightens even as what it holds grows abstract. You see this already beginning to happen when a square number is distinguished from a square shape.

The Pythagoreans wanted to find all triples of whole numbers, (a, b, c), such that $a^2 + b^2 = c^2$. They knew that given one such, all multiples of it, (ma, mb, mc), would be triples too, so their search narrowed down to finding the ancestors of each family line: those triples that had no common factor (or were 'relatively prime', as they say in the trade). This is the time to sit under your favorite tree and close your eyes.

On reflection, we can be sure that if any three numbers satisfying our equation have no common factor, no two of then can share a factor either. For if a and b, for example, were both multiples of some prime p, then $a = pk$ and $b = pm$ for some k and m. But since $a^2 + b^2 = c^2$, this would mean that

$$(pk)^2 + (pm)^2 = c^2$$

$$\Rightarrow p^2(k^2 + m^2) = c^2$$

which means that c also has p as a common factor, and we agreed that the triple (a, b, c) had no factor in common.

Very well: we're therefore looking for ancestors in which each pair is relatively prime. It would have been natural for a Pythagorean to think at this point in terms of *parity*: whether a number is odd or even (since this distinction awoke so many hidden harmonies for them). They would therefore rephrase our latest conclusion by saying that a, b, and c can't all be even, nor could any pair of them be—hence all three must be odd, or two will have to be odd and one even. This is the sort of dividing and conquering that number theory* delights in. It was how Socrates caught the sophist in his nets,[1] and how we, along with those ancient mathematicians, will catch the infinite variety of Pythagorean triples in a little box.

It can't be that a, b, and c are all odd, since an odd number is of the form $2s+1$, and so an odd squared is odd. But this means that if a, b, and c were all odd, $a^2+b^2=c^2$ would also be composed of all odd numbers. Yet an odd plus an odd is even, so we would have a^2+b^2 (an even) equal to c^2 (an odd), which is impossible. Hence *two* of a, b, c are odd and one is even. Which?

Were c even, say $c=2m$, then $c^2=4m^2$, and 4 would be a divisor of c^2. But since a and b have to be odd, we can write $a=2p+1$ and $b=2q+1$ for some p and q. That makes $a^2+b^2=(2p+1)^2+(2q+1)^2=4(p^2+p+q^2+q)+2$. Hence 4 can't be a divisor of a^2+b^2 (since it is a factor of the first but not the second of the terms that make it up), which contradicts the fact that it divides $4m^2=c^2$, which these add up to. So c can't be even, and since we must have c odd, that means either a or b must be even. Does it matter which? No. Let's take a to be odd and b even.

Pause to appreciate how far we've come. Invoking no more than parity, we've learned something important about a, b, and c—their genetic makeup, if you will. For what it's worth, since c and a are both odd, $c+a$ and $c-a$ must both be even.

Let's make use of this little knowledge by saying in a different way the only other thing we know: $a^2+b^2=c^2$ means that $b^2=c^2-a^2$, which in turn equals $(c+a)(c-a)$. Yes, this confirms that b is even (since even times even is even), but it will turn out to do much more. For we claim that the *only* common factor of $(c+a)$ and $(c-a)$ is 2. While we have more or less followed our noses up to now, this latest claim isn't at all obvious.

---

* Number theory is that branch of mathematics which, like life, raises questions so easy to ask and so hard to answer. More specifically, the questions it asks are about the properties of whole numbers.

Here's the ingenious proof (it's like watching a major league short-stop turn a double play on a bad hop that would likely skip past a fielder in the minors). If $c+a$ and $c-a$ had in common a factor other than 2, call it d, then $c+a=2df$ and $c-a=2dg$, for some whole numbers f and g. Dividing by 2, we would then have:

$$\frac{(c+a)}{2} = df, \qquad \frac{(c-a)}{2} = dg.$$

Adding these two together we get

$$
\begin{array}{r}
\dfrac{c+a}{2} = df \\[2mm]
+ \quad\quad\quad \\[-1mm]
\dfrac{c-a}{2} = dg \\[1mm]
\hline
\dfrac{2c}{2} = df + dg \\[2mm]
c = d(f+g)
\end{array}
\qquad (1)
$$

This means that d is a factor of c.

Now let's *subtract* our second term from the first:

$$
\begin{array}{r}
\dfrac{c+a}{2} = df \\[2mm]
- \quad\quad\quad \\[-1mm]
\dfrac{c-a}{2} = dg \\[1mm]
\hline
\dfrac{2a}{2} = df - dg \\[2mm]
a = d(f-g)
\end{array}
\qquad (2)
$$

This means that d is a factor of a. But that is impossible: c and a can't have a common factor—they are relatively prime.

Hence when we divide $(c+a)$ and $(c-a)$ by 2, the results are relatively prime.

Let's assess our gains: a and c are odd, b, $(c+a)$ and $(c-a)$ are even (which means that $\dfrac{b}{2}, \dfrac{(c+a)}{2}$ and $\dfrac{(c-a)}{2}$ are whole numbers), and $\dfrac{(c+a)}{2}$ and $\dfrac{(c-a)}{2}$ are relatively prime.

Recalling that $b^2 = c^2 - a^2 = (c+a)(c-a)$, let's divide through by 4 (hindsight alone would have justified this seemingly random step):

$$\frac{b^2}{4} = \frac{(c+a)(c-a)}{4} \Rightarrow \left(\frac{b}{2}\right)^2 = \frac{(c+a)}{2} \cdot \frac{(c-a)}{2}. \qquad (3)$$

But since $\dfrac{(c+a)}{2}$ and $\dfrac{(c-a)}{2}$ are relatively prime, each is an integer squared!

Why? Because their product $\left(\dfrac{b}{2}\right)^2$ is a square; since the terms $\dfrac{(c+a)}{2}$ and $\dfrac{(c-a)}{2}$ have no common factor, that means there is no number n in both of them which, when they are multiplied together, would make an n^2. So these two factors must, independently, be squares.

Let's say

$$\frac{(c+a)}{2} = u^2 \text{ and } \frac{(c-a)}{2} = v^2. \qquad (4)$$

It may seem that this dog has been chasing its own tail, but in fact he has just dug up the bone. Watch. Replacing the right-hand side of equation (3) above,

$$\left(\frac{b}{2}\right)^2 = \frac{(c+a)}{2} \cdot \frac{(c-a)}{2}.$$

With our new names, we find that $\left(\dfrac{b}{2}\right)^2 = u^2 v^2$. So

$$\frac{b^2}{4} = u^2 v^2 \text{ and}$$

$$b = 2uv.$$

Adding the two equations in (4) gives us

$$c = u^2 + v^2,$$

while subtracting the second from the first in (4) tells us that

$$a = u^2 - v^2.$$

Since $\dfrac{(c+a)}{2}$ and $\dfrac{(c-a)}{2}$ are relatively prime, so are u^2 and v^2, hence so are u and v. Finally, since c and a are both odd, one of u^2 and v^2 is even and one is odd (were both odd, their sum and difference would both be even, and $\dfrac{(c+a)}{2}$ and $\dfrac{(c-a)}{2}$ wouldn't be relatively prime).

What does this mean for our quest? If you choose any two relatively prime whole numbers, u and v—one even, one odd (and let's assume u is greater than v)—then you will *automatically* have made a Pythagorean triple:

$$a = u^2 - v^2$$
$$b = 2uv$$
$$c = u^2 + v^2.$$

Does $a^2 + b^2 = c^2$, for these values? It's the work of a minute to check: and yes indeed, this compact little kit yields all and only the primitive ancestors of the whole Pythagorean family. If you wish to dot your i's and cross your t's by confirming that when a, b, and c are so chosen they really are relatively prime, see the appendix.

This outcome is up there with a fugue from *The Well-Tempered Clavier* or the dome of St. Paul's. By pointedly looking away from the visible world, we have seen not only the two families of gnomonic triples discovered in Chapter Two, but triples you wouldn't have guessed, in a genealogy limitlessly broad. Here, for example, are the choices of u and v for some members of our first (single gnomon) family. If $u=2$ and $v=1$, we get (3, 4, 5); $u=3$ and $v=2$ produces (5, 12, 13). From the second (two-gnomon) family, let $u=4$ and $v=1$ to get (8, 15, 17). Do you recall Neugebauer's triple (4601, 4800, 6649)? It comes from $u=75$ and $v=32$.

Choose your own monstrously large u and v fitting the requirements: you will have just made a Pythagorean triple most likely never imagined before.

Should you conclude that so much doggedness, wedded to such inspiration, is beyond your wildest dreams, remember that what you've followed here is the tidied remainder of who knows how many lively conversations—and conversations among practitioners as devoted to their craft as are cooks to theirs. Is a pinch of cardamom, a drop of vanilla, a hint of lemon zest needed to make the dish just so? And here: is it factoring, or thinking in terms of parity, or of primes, that would perfectly season our thinking? Construing then, as now, evolves past constructing, and accounts for how mathematics creates what it later reads as discovered.* This is why Einstein could say: "I hold it true that pure thought can grasp reality, as the ancients dreamed."[2]

△ △ △

What we've seen, in the previous chapter and this, has been not only proofs but a short history of proving. Let's deepen that history by comparing this Greek way of finding all Pythagorean triples to its modern counterpart.† Just as tautologies let us look below the surface by saying the same thing in two different ways, so contrasted proofs of the same theorem tell us a great deal about the nature of shifted points of view. They might even raise in your mind the question of whether a theorem remains the same, when differently proved. Context alters content.‡

---

* The art of construing, as you've seen it practiced here, involves reworking mathematical sentences—i.e., equations. No wonder it might seem like hollow tautological play to the Hearers—but to the Knowers? It shifts massive concepts by moving the symbols that are their visible levers.

† Diophantus seems to have come up with a very similar algebraic method. Some think of him as a modern mathematician who just happened to live two thousand years ago; others, as one of those exemplars of pioneering, whose baffling writings are unexplored seas of their own.

‡ They might also set you wondering whether things are differently true on the whole than specifically because truth is a formal property, while

Since we're looking for all triples (a, b, c) of natural numbers such that $a^2 + b^2 = c^2$, and since c, being a natural number, can't be zero, we could divide by it and rewrite:

$$\left(\frac{a}{c}\right)^2 + \left(\frac{b}{c}\right)^2 = 1.$$

Think of this equation as saying that we're looking for all pairs of rationals, a/c and b/c, which solve the equation

$$x^2 + y^2 = 1 \tag{1}$$

Rather than fleeing from geometry to number, as we did above, let's call geometry to our aid: this is the equation of a circle with radius 1,

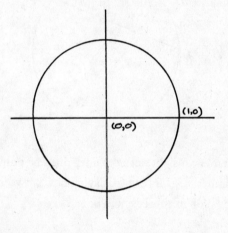

imagination seems to grapple with and ultimately grasp the specific. Or is imagination a messenger, born of that interplay between example and exemplar, which has been a theme running through this book? What does the drawing of a triangle you see on the page stand for: its shaky self, with all the accidents of ink and position; triangularity per se; something beyond—or in between? Shakespeare's prologue to *Henry V* asks you to

> Think when we talk of horses, that you see them
> Printing their proud hoofs i' the receiving earth;
> For 'tis your thoughts that now must deck our kings.

Imagination both bodies forth and disembodies.

centered at the origin of the Cartesian plane. What we're asking for is all points on this circle whose coordinates, (x, y), are both rational.

We can quickly come up with four candidates: $(1, 0)$, $(0, 1)$, $(-1, 0)$, and $(0, -1)$. Are there others—and if so, how shall we find any or all of them? Half the art of invention lies in choosing a fruitful viewpoint—the other, in seeing something useful from it. While we know ourselves too little yet to be sure where such insights live, it often helps to ring a few doorbells. Geometry, circles, points . . . what about lines? Let's anchor a straight line at one of these points—say $(-1, 0)$—and let it sweep its way across the circle:

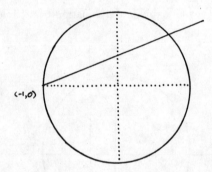

Second, paired, insight: let's consider this line only when its slope is a rational number m—for then (parlaying now between geometry and algebra) the equation of this line will be

$$y = mx + b,$$

and since $(-1, 0)$ is on this line—i.e., $x = -1$, $y = 0$—we can solve for b and find $b = m$, so that our equation

$$0 = m(-1) + b$$

turns $y = mx + b$
into $y = mx + m,$

or

$$y = m(x+1). \qquad (2)$$

We've set it up so that if x is a rational number, so is y; and if these points (x, y) lie both on the line and the circle—that is, where the line meets the circle a second time, after its intersection at $(-1, 0)$—then those points will be the ones we want: points with rational coordinates $x = \dfrac{a}{c}$, $y = \dfrac{b}{c}$ satisfying $a^2 + b^2 = c^2$.

It remains only to find them—and this needs not art but craft: the craft of handling equations in general, and quadratic equations in particular. We want to solve our equations simultaneously, so taking out your dividers and multipliers from their velvet cases and substituting (2) in $x^2 + y^2 = 1$, you find

$$x^2 + (m(x+1))^2 = 1,$$

so

$$x^2 + m^2 (x^2 + 2x + 1) = 1.$$

Write this all out—

$$x^2 + m^2 x^2 + 2m^2 x + m^2 = 1,$$

sort it—

$$x^2(1 + m^2) + 2m^2 x + (m^2 - 1) = 0$$

for the final tidying:

$$(m^2 + 1)x^2 + 2m^2 x + (m^2 - 1) = 0. \qquad (3)$$

An easy way now to find all possible x (from which, by using equation (2), we'll then get the paired possible y) is this. Since we know $x = -1$ is a solution, i.e., $x + 1 = 0$, let's divide (3) by $x + 1$ to get the other root:

$$x + 1 \overline{)(m^2 + 1)x^2 + 2m^2x + (m^2 - 1)} \atop (m^2 + 1)x + (m^2 - 1)$$

(If you don't want to check this by doing the division, just multiply the quotient by the divisor to see if you get the dividend back.)

The other root is therefore the solution of

$$(m^2 + 1)x + (m^2 - 1) = 0 \qquad (4)$$

i.e.,

$$(m^2 + 1)x = 1 - m^2,$$

so

$$x = \frac{1 - m^2}{1 + m^2} \qquad (5)$$

Substituting that value for x into our equation (2) for y,

$$y = m \cdot \frac{1 - m^2}{1 + m^2} + 1$$

$$y = \frac{m(1 - m^2) + (1 + m^2)}{1 + m^2}$$

$$y = \frac{2m}{1 + m^2}$$

This means that for every rational number m we get a solution to (x, y) in rationals:

$$x = \frac{1 - m^2}{1 + m^2} \quad y = \frac{2m}{1 + m^2} \tag{6}$$

We've managed to show that if m is rational, then we have a solution in rationals to (1). But are these *all* the rational solutions? That is, if (r, s) is a solution in rationals to (1), will there be an m that produces it? Well, if (r, s) is such a solution, then the line through it and $(-1, 0)$ will have slope

$$m = \frac{s - 0}{r + 1} = \frac{s}{r + 1},$$

which is rational. Our modern process has indeed yielded *every* rational solution to (1).*

How very different the results of the Greek and modern approaches look. Has the nature of number, or of the Pythagorean relation, changed over the millennia, or need we only compare? Let's see. If we write our rational m as $m = \dfrac{v}{u}$ and substitute this in (6), we have

$$x = \frac{u^2 - v^2}{u^2 + v^2} \text{ and } y = \frac{2uv}{u^2 + v^2}.$$

In other words, $\dfrac{a}{c} = \dfrac{u^2 - v^2}{u^2 + v^2}$, $\dfrac{b}{c} = \dfrac{2uv}{u^2 + v^2}$, so once again, $a^2 = u^2 - v^2$, $b = 2uv$ and $c = u^2 + v^2$.

One difference, though, is that in these liberal times our procedure has given us all, not just the ancestral, Pythagorean triples. It even includes a triangle with sides $-3/5$ and $-4/5$ and hypotenuse 1—not a generalization even our most adventurous travels came upon. To restrict ourselves to the ancestors we would need to put the old restraints back on u and v: that they be relatively prime whole numbers, one even, one odd, and with $u > v$.

---

* The only point with rational coordinates on the circle which our approach won't capture is $(-1, 0)$ since the line through it is vertical, and hence has an infinite slope m.

We've been describing the modern counterpart of the way the Greeks handled Pythagorean triples—but "modern" is a floating index. The Sumerians must have felt as modern as we do (recall, for example, that the Hypermodern school of thought in chess flourished a century ago). Since there never seems to be a last word in mathematics, would it surprise you to hear of this leap past the steps we've just taken: c is the hypotenuse of an ancestral Pythagorean triple (a, b, c) if and only if it is a product of primes, each of which leaves a remainder of 1 when divided by 4?[3]

△ △ △

The dynamic character of contemporary approaches may well have struck you. What has become of the Greek conviction that things mathematical are immovably part of what *is*? A great many mathematicians still consider themselves Platonists (a name that covers many variations), but an altered view of mathematical activity—from constructing to construing, as we've put it—allows them to shift and make shifts in the middle world of *language*: processes at whose limit those immutable structures lie. The fecund tension is no longer between geometry and algebra, but *play* within and between them generates insightful ways of re-saying and so re-seeing the eternal same, as mind planes over the procreating world.

You catch a flicker of this evolution in the movement from Loomis's old strictures against trigonometric proofs of the Pythagorean Theorem, to the trigonometry-like work we've just done together in finding Pythagorean triples. The One remains, the terms it is that change.

APPENDIX

The fine print guaranteeing that our hard-won a, b, and c are indeed relatively prime reads like this:

Since a and c are both odd, 2 couldn't be a common factor. But say they had an odd common factor, p, so that $a = pm$ and $c = pk$. Then since

$$\frac{(c + a)}{2} = u^2$$
$$2u^2 = c + a = p(k + m)$$

and since

$$\frac{(c - a)}{2} = v^2$$
$$2v^2 = c - a = p(k - m).$$

This means p would be a factor of both $2u^2$ and $2v^2$, hence of u^2 and v^2—and therefore of u and v.

But these, we said, were chosen to be relatively prime—so, indeed, a and c have no common factor.

Had a and b a common odd factor q, since $b = 2uv$, q would have to be a factor of u or v—say u. But $a = u^2 - v^2 = (u + v)(u - v)$, hence q must divide one of these terms. But since it divides u, it would also therefore have to divide v, which is impossible, since u and v are relatively prime.

CHAPTER EIGHT

Living at the Limit

We've traveled very far since leaving a bold conjecture stranded, half clothed, in Chapter Six: might the square on the 'body-diagonal' of an n-dimensional box be the sum of the squares on the n sides of the box? We know this is true for $n=2$ (for this is the Pythagorean Theorem), and saw it was true for $n=3$.

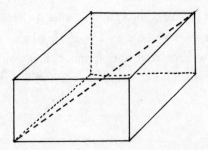

Being experienced voyagers, we are confident that induction will easily carry us home.* For assuming the conjecture true in k dimensions, the square on the body-diagonal of a $k+1$ dimensional box will (in the plane made by the new side in dimension $k+1$ and the previous body-diagonal in dimension k) be the square of the new side plus the square on that previous diagonal, by the Pythagorean Theorem; and by the inductive assumption, this square is itself the sum of the squares on the previous k sides. When Descartes, as you saw, wrote 1, 2, 3, 4 in his

---

* We'd cast off on induction (in the appendix to Chapter Six) as showing us only *that*, not *why*, a statement was true. But it has this compensatory virtue: it takes our finite insights to infinity, where perhaps *why* and *that* converge.

secret journal, he might have whispered in his mind: ". . . "—where the dots bear infinitely more weight than the sturdy integers that precede them.

FROM HERE TO ETERNITY

Two further journeys open from this: one will change our viewpoint, the other immeasurably broaden it. Both begin by acknowledging that the n-dimensional box is a natural framework for the Pythagorean Theorem, with the hypotenuse as a diagonal strut. But aren't we perversely making nested matryoshka dolls? Why picture such a box *in* n-dimensional space, when that space could itself be all the framework we require? For it is a 'box' open in every direction from the neat corner at the origin where its n axes meet, each at right angles to the rest.

One of us, at thirteen, got the shock of a lifetime on reading in Sir Arthur Eddington's little book, *The Expanding Universe*, his explanation of why the galaxies were all receding from ours. Think of the galaxies, he wrote, as ink drops spattered on a balloon, which you then inflate. Each drop sees the others as moving away from it, but in fact it is the space itself which is expanding. So that's how adults thought of things! Well, perhaps not—but it is the view geometers come to: what they study isn't so much shapes in space as the space of shapes. What we'll do now is rethink the Pythagorean Theorem as a property of n-dimensional space.

For this we'll need such things as angles with their measures, and lines along with their lengths, and the squares of these lengths; none of these imported from elsewhere, but (in the spirit of trying to reach the innermost doll) intrinsic. To that end, we won't think of space as a grid of points with coordinates, but as a *vector space:* a set of line segments each with a direction and length, most naturally pictured as arrows with their tails at that common origin and their heads at what before were just those isolated points.

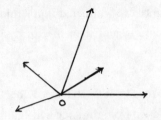

How easily now a limpid picture begins to coalesce. That origin, as with everything else in this space, is itself a vector—whose length is 0. Out of it, like quills upon the fretful porpentine, grows a set of special arrows, each at right angles to the others, one for each of the space's *n* dimensions—and again, for the simplicity that uniformity brings, let each of these be of length 1. These special vectors are the *axes* for the space, and the set of them is aptly called a *basis*. In two dimensions therefore the basis consists of the two vectors {(1, 0), (0, 1)}; in three, {(1, 0, 0), (0, 1, 0), (0, 0, 1)}—and so on. As a shorthand for the *n* axes in *n*-dimensional space, these vectors tend to be called $e_1, e_2, \ldots, e_n$.

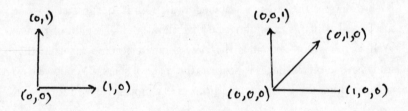

In order to stretch or shrink these axes to an infinity of other vectors in their directions, we provide ourselves with a kit of *scalars*: all the real numbers, thought of as multipliers.* To make a vector 17 units long in the e_1 direction, for example, just multiply each coordinate of e_1 by 17: $17e_1 = (17, 0, 0, \ldots, 0)$.

And what about all the vectors that don't point along any axis but are at angles to them? We can form these with one of the last bits of

---

* As you can imagine, we could pack our kit with complex numbers instead, or fancier sorts of multipliers—so long as each kit had a 0 in it, and the additive inverse of anything else it contained.

machinery we'll need. For we want to be able to add vectors to one another, as well as multiply each by scalars. But how do you add two arrows? As so often happens, physics gives us the clue: if you have two component forces acting on a particle, it squirts out between them in a quite specific way. Picture this in two dimensions: the resultant of the component forces (vectors) X and Y is their sum, X+Y, understood as the long diagonal of the parallelogram these components generate:

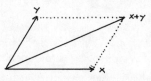

To speak of this as a 'sum' isn't as arbitrary as it may at first seem, because the coordinates of the vector X+Y are indeed the sums of the respective coordinates of X and Y: $(x_1 + y_1, x_2 + y_2)$.

So the vector that is the sum of X = (1, 0) and Y = (0, 1) is X+Y = (1, 1):

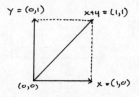

and (staying in two dimensions for ease in visualizing), you could write a vector like (8, −1) as (2, 2) + (6, −3). Or at an even further remove, since (2, 2) is 2(1, 1) and (6, −3) is 3(2, −1), why not write (8, −1) = 2(1, 1) + 3(2, −1).

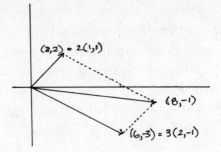

In this way we can manufacture any vector we like from the sum of others, suitably scaled.

The only work left to do before rephrasing the Pythagorean Theorem in the language of vector spaces is to understand how to calculate a vector's length, and what signals that two vectors are perpendicular to each other. We'll let the Pythagorean Theorem itself lead us. The vector $(4, 0)$ has length $\sqrt{4^2 + 0^2} = 4$; the vector $(0, 3)$ has length $\sqrt{0^2 + 3^2} = 3$, and the vector $(4, 3) = (4, 0) + (0, 3)$ has length $5 = \sqrt{4^2 + 3^2}$: that is,

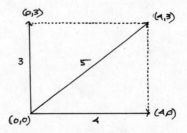

the square root of the sum of each of its coordinates squared. This accords with our common intuition of distance, and is what the rectangular box on page 192 showed us: The length of a vector (x, y) in two dimensions is $\sqrt{x^2 + y^2}$, and of a vector $(x_1, \ldots, x_n)$ in n dimensions is $\sqrt{x_1^2 + \cdots + x_n^2}$.

So, fancying up the absolute value sign commonly used for distance, we write the length of a vector $X = (x_1, \ldots, x_n)$ as

$$\|X\| = \sqrt{x_1^2 + \cdots + x_n^2}.$$

To go along with this elevated notation, the custom is to rechristen this more abstract length 'norm'.

All has been just as it should and had to be. What is surprising now, however, is to take this operation (multiplying of a vector's coordinates by themselves and then adding them all together) as a special case—with one vector—of what we did above with two vectors, $X = (x_1, \ldots, x_n)$ and $Y = (y_1, \ldots, y_n)$:

$$x_1 y_1 + x_2 y_2 + \cdots + x_n y_n.$$

In keeping with that innermost doll we're after, this expression is called the 'inner product' of the vectors X and Y, and is represented by $\langle X, Y \rangle$:

$$\langle X, Y \rangle = x_1 y_1 + x_2 y_2 + \cdots + x_n y_n.$$

Notice that this inner product of two vectors isn't a vector, but a *number*—a scalar, in the setting of vector spaces.

Wonderfully enough, it will let us recognize when two vectors are perpendicular. Take $X = (4, 0)$ and $Y = (0, 3)$:

$$\langle X, Y \rangle = 4 \cdot 0 + 0 \cdot 3 = 0.$$

What about another pair of perpendiculars, $X = (1, -2)$ and $Y = (4, 2)$?

$$\langle X, Y \rangle = 1 \cdot 4 + -2 \cdot 2 = 0.$$

Could it be that two vectors X and Y are perpendicular if and only if their inner product is 0? Yes—and in fact its proof sped by in Chapter Six, disguised as the Law of Cosines! You will find the details worked out in this chapter's appendix.

Looking more broadly, we see a space of vectors (with attendant scalars) and an inner product defined on it, which lets us calculate length and judge perpendicularity. For any vector X in n-dimensional vector space, $\|X\| = \sqrt{\langle X, X \rangle^2}$, and $X \perp Y \Leftrightarrow \langle X, Y \rangle = 0$. Peculiar as they look, inner products seem to unlock inner doors.

We have, therefore, for any two perpendicular vectors X and Y:

$$\|X\|^2 + \|Y\|^2 = \|X + Y\|^2.$$

This holds in any n-dimensional vector space for a set of mutually perpendicular vectors:

$$\left\| \sum_{i=1}^{n} X_i \right\|^2 = \sum_{i=1}^{n} \|X_i\|^2.$$

A Dijkstra dream of elegance has dawned: "The Pythagorean Theorem? Oh, you mean 'for mutually perpendicular vectors, the squared norm of the sum equals the sum of the squared norms.'" Surely, though, we have here something more substantial than elegance—a clarifying effect on our rambling knowledge which isn't decorative but structural. For we have found that we aren't all at sea in n-space, but can always navigate just from a planar map within it. You saw this in the context of the n-dimensional box: the length of the new hypotenuse in $(n+1)$-dimensional space is calculated from the new perpendicular in that space and the old hypotenuse in n-space: on that two-dimensional plane, therefore.

△ △ △

And can this view of space via vectors bring us now to that second voyage as well, and a continent hardly dreamed of in our milling about the close streets of the old capital? Past the next dimension and the next and the next after that, their sequence stretching endlessly away, might there be a space which was *in itself* infinite-dimensional? The great accommodating generality of vector spaces will let us see that there is; will let us grasp what finite needs of ours drive us to it; how we should imagine it, and ourselves in it; and how we are to find familiar shores there— such as the Pythagorean Theorem. You might think that such plans would materialize only in the gauziest panoramas—

All the most childish things of which idylls are made[1]

—in fact it is an ultimate *accuracy*, possible only in infinite dimensional space, to which the lust for ever improving *approximation* will lead us.

You saw in Chapter Five (p. 88) how adding higher and higher powers of x (each with the right coefficient) more and more closely matched the value of the sine function at a specific point. The whole of the infinite power series would give its value at that point precisely. Even with all its terms, however, the power series representation of a function, at a point, fits it less and less well as you move away from that point. Here you see several terms of the power series for sin x, at $x=0$, peeling away from it as we move off from 0.

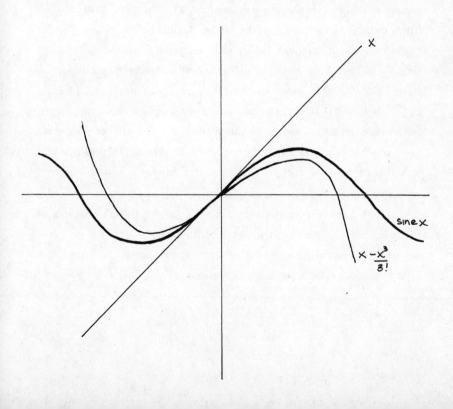

The urge from physics to analyze complex motion (heat diffusing across metal, waves crashing on the shore) into simple components led to a search for how to replace any function by a sum of more basic ones, not just at a point but over a suitable stretch. Some two hundred years ago in France, Joseph Fourier (a tailor's son who became a mathematician, as well as Napoleon's governor of Lower Egypt) stitched together a way to do this. As a result, *Fourier series*—no more than sines and cosines of various frequencies and amplitudes, added up—better and better match the required function over an interval as more and more terms appear, and match it *exactly* when the sum is infinite. The choice of trigonometric functions for these threads may strike you as peculiar, but the oddity is lessened if you've ever played with an oscilloscope, or seen in science fiction or hospital dramas those green waves coalescing into a straight line, since oscilloscopes make Fourier series visible.

Very real demands, therefore (from wavering bridges to magnetic resonance imaging)—not only a dream beyond induction—have led us to infinite-dimensional vector spaces, whose flexible neutrality gives us the perfect way to understand, build, and manipulate these wonderful series. Why infinite-dimensional? Because—while the vectors will now be functions, $f(x)$, $g(x)$, etc.—the axes e_1, e_2, ... will no longer be $(1, 0, 0, \ldots, 0)$, $(0, 1, 0, 0, \ldots, 0)$ and so on, but those basic components, $\sin mx$ and $\cos nx$, for each of the infinitely many natural numbers m and n.

To set this up requires, as before, an inner product $\langle f, g \rangle$ which will guarantee that these axes are mutually perpendicular, and will also let us derive a norm from it, just as in the finite case: $\|f\| = \sqrt{\langle f, f \rangle^2}$. We needn't fetch far: the inner product for functions will be analogous to the one we had. See this chapter's appendix for details.

To see how quickly a Fourier series converges to the function it represents, let's take $f(x) = x$, whose series is

$$2\left(\sin x - \frac{\sin 2x}{2} + \frac{\sin 3x}{3} - \frac{\sin 4x}{4} + \cdots \right)$$

and look at graphs with 1, 3, 5, and 10 terms:

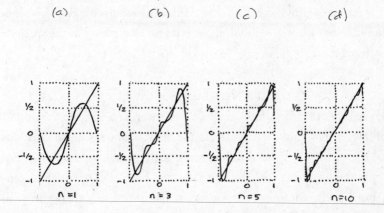

The eye takes this in more quickly than the mind struggling with notation. The ear is even faster: we hear by breaking the sound wave $f(x)$ that strikes the eardrum into the terms of such series, and find that the coefficients in these terms measure the sound's unique qualities. A sound is familiar if it has the same Fourier coefficients as another we have heard. An inner product indeed.

We now have our infinite dimensional vector space, and the instruments for steering our way through it. These lead us at once to the exact analogy of the Pythagorean Theorem in n-dimensional space: just as there, if we have a set $\{X_i\}$ of mutually perpendicular vectors, then

$$\left\| \sum_{i=1}^{\infty} X_i \right\|^2 = \sum_{i=1}^{\infty} \|X_i\|^2,$$

so long as the infinite sum on the right converges.

As everywhere in math, the *form* remains, however various the content that takes it on. This latest incarnation of the Pythagorean Theorem is called Parseval's Identity, after Marc-Antoine Parseval, who—twenty years before Fourier—was thinking in terms of summing infinite series, without Fourier's specifics in mind. He is remembered as a French nobleman who survived the Terror, wrote poetry against Napoleon, failed (even after five attempts) to be elected to the Académie des sciences, and was eccentric—in unspecified ways. This equation having been his identity may explain why his mortal one remains as shadowy as the space he pictured it in.

DISTANCE AND CLARITY

If you think that the infinity we have been savoring belongs to the ivoriest of towers, look about you: we live in the midst of limits. Our speculations ever speciate, our generalizations endlessly generate, so that a whole is grasped only through the vision of a horizon. Each of us stands at the center of a circle of infinite radius (since, as Pascal first saw, that center will be everywhere). This is no idle metaphysics: in a world founded on the straight and the rectangular, where like Masons we all feign living on the Square, the paths we make are curved and our expectations jagged. Yet errors average out, the curves are sums of infinitely many infinitesimally small straight lines, and the Pythagorean Theorem holds the key to understanding both.

Suppose that a curve like this

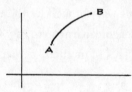

is the graph of a smooth function $f(x)$ from point $A = (a, f(a))$ to $B = (b, f(b))$.

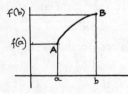

What is its length l? The roughest approximation l_1 would be the length of the straight line joining its end-points:

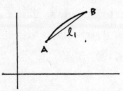

and by the Pythagorean Theorem,

$$l_1 = \sqrt{[f(b) - f(a)]^2 + (b - a)^2}.$$

A better approximation, l_2, breaks the straight line in two,

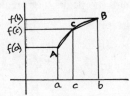

and $l_2 = \sqrt{[f(c) - f(a)]^2 + (c - a)^2} + \sqrt{[f(b) - f(c)]^2 + (b - c)^2}.$

Keep subdividing the interval from a to b, getting more and more successively shorter segments, each the hypotenuse h of a triangle with base Δx and height Δy. Here is a typical one:

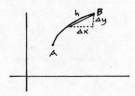

So the length of the curve is approximately

$$l_{\text{approx}} = \sum_{k=1}^{n} \sqrt{(\Delta x_k)^2 + (\Delta y_k)^2}.$$

But it is the *exact* length we're after: the limit, therefore, of the sum of these hypotenuse secants, each touching the curve not at two points (no matter how close) but ultimately so short as to touch it only at one— not secants anymore, that is, but *tangents*.

We want something like

$$l = \lim_{n \to \infty} \sum_{k=1}^{n} \sqrt{(\Delta x_k)^2 + (\Delta y_k)^2}, \tag{1}$$

with the sides Δx_k and Δy_k diminished to 0.

Here differential calculus comes to our rescue: that application of the old Scottish proverb "Mony a mickle maks a muckle." It guarantees that a curve smooth enough (no peaks ∧ or breaks ∼ ∼) to have a slope at every point on it (if the graph is of $f(x)$, these slopes are read off by the *derivative* function, $f'(x)$) will have at least one point p_k in every interval Δx_k, where the slope of the graph is the same as the slope of the hypotenuse secant:

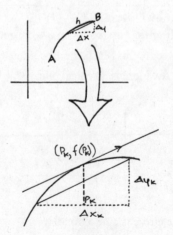

$$f'(p_k) = \frac{\Delta y_k}{\Delta x_k}, \text{ so } \Delta y_k = f'(p_k) \cdot \Delta x_k.$$

This means we can rewrite (1) this way:

$$l = \lim_{n \to \infty} \sum_{k=1}^{n} \sqrt{(\Delta x_k)^2 + (f'(p_k) \cdot \Delta x_k)^2}$$

$$= \lim_{n \to \infty} \sum_{k=1}^{n} \sqrt{1 + (f'(p_k))^2} \cdot \Delta x_k. \tag{2}$$

But we want not only infinitely many of these Δx_k: we want the width of each to shrink to 0 (which will also constrain the p_k in it to a single x):

$$\lim_{\Delta x_k \to 0} \lim_{n \to \infty} \sum_{k=1}^{n} \sqrt{1 + (f'(x_k))^2} \, \Delta x_k$$

Now integral calculus seizes on this clumsy expression and recognizes it as the sum from a to b of the 'areas' of trapezoids with height $\sqrt{1 + (f'(x))^2}$ and infinitesimal width dx—so that the length of the curve becomes exactly

$$l = \int_a^b \sqrt{1 + (f'(x))^2} \, dx \overset{*}{.}$$

Thus the curves of life are each the living limit of yet another avatar of the Pythagorean Theorem, whose image multiplies and miniaturizes and so more fully pervades thought, the closer we look.

△ △ △

Why is the Pythagorean Theorem everywhere? Because life is movement, movement begets measurement, and we measure distance along shortest paths, which the Theorem gives us. Must it be so? Not really: anything will serve to measure $d(X, Y)$, the distance between two points X and Y, if it yields positive values (and 0 only if X is Y); makes

---

* Much of this explanation is in a kind of shorthand that abbreviates centuries of engineering and numerous careful manoeuvres with inequalities, to ensure that such arrows of desire as in $n \to \infty$ and $\Delta x_k \to 0$ actually fall past any specified boundary for the one, and within any required ring from the bull's-eye for the other.

the distance from here to there the same as that from there to here—
$d(X, Y) = d(Y, X)$; and which makes $d(X, Y)$ the shortest length between them: that is, $d(X, Y) \leq d(X, Z) + d(Y, Z)$, with equality only
if Z is on the line XY.

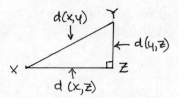

This 'triangle inequality' is the signature on the guarantee that you have
bought a genuine measurer of distance.

Here are a few examples of other walnut and brass measuring
devices:

Let $d(X, Y)$ be 0 if $X = Y$ and 1 if $X \neq Y$. This satisfies all three of
our requirements (though it does seem a bit odd that it makes all non-
zero distances the same).

If we stick to a line, $d(X, Y) = |X - Y|$ works perfectly well, as does
$d(X, Y) = \dfrac{|X - Y|}{(1 + |X - Y|)}$. This last has the peculiar feature that all distances will be less than or equal to 1.

Why not use one of these, or some other measure? Partly because
some are too coarse; mostly because the Pythagorean way of measuring
is so deeply natural (as long as we act as if we lived on a plane): if
$X = (x_1, x_2)$ and $Y = (y_1, y_2)$, then

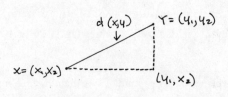

$$d(X, Y) = \sqrt{(y_1 - x_1)^2 + (y_2 - x_2)^2},$$

which generalizes nicely to 3- or n-dimensional Euclidean space:

$$d(X, Y) = \sqrt{(y_1 - x_1)^n + \cdots + (y_n - x_n)^n}\,.^*$$

We become so used to measuring distance by the Theorem that we forget how awesome are the paths of pursuit. The third baseman and shortstop trap their swerving opponent on the shrinking diagonal he takes between them. The predators on the veldt that ran right at their moving targets are extinct: those survived who ran to where their inner Pythagoras calculated the prey was going.[†]

But even if we grant that the Theorem dictates trajectories in physical space, do we recognize its power when we wonder what direction our lives should take? Longing can perplex us, as it did the war poet Henry Reed:

> . . . I may not have got
> The knack of judging a distance; I will only venture
> A guess that perhaps between me and the apparent lovers
> . . . is roughly a distance
> Of about one year and a half.[2]

We weigh up the odds; we try to come to terms with the random residues of our best-laid plans, and the errors that swarm like mosquitoes around our every accuracy. Looking for some way to draw order out

---

[*] If we look at the even more general measure

$$d(X, Y) = [(y_1 - x_1)^p + \cdots + (y_n - x_n)^p]^{1/p},$$

we see that the absolute value measure is just the special case $p = 1$. For just about all of these measures, it takes some fancy algebraic dancing to prove the triangle inequality.

[†] Second order tactics, of course, such as stealth and deception, can modify the most Pythagorean of strategies. So the shortest distance in proof space is from insight to confirmation, but the richest voyages include those detours to sudden views that open up speculation on other and deeper vistas.

of data, we naturally cast it in the Pythagorean mold. Given a set of numerical readings, $X = \{x_1, x_2, \ldots, x_n\}$, statisticians find its mean, M,

$$M = \frac{(x_1 + x_2 + \cdots + x_n)}{n}$$

to lend some stability to these points, then measure with the Theorem how far each datum is from this mean (there's that n-dimensional form of the Theorem again); average out the measurements; and so find the taming 'standard deviation', $\sigma(X)$:

$$\sigma(X) = \frac{\sqrt{((x_1 - M)^2 + (x_2 - M)^2 + \cdots + (x_n - M)^2)}}{n}.$$

This gives some sort of shape to chaos. Pythagoras has thus returned in his original role as a seer—by reading not entrails but the detritus of the day.*

We live under the Theorem's guidance not only in a landscape of curves at the limit of approximating chords, but at the one and only limit of the likely: the actual world as it hurtles through phase space (defined back in Chapter Six, p. 164). This is where we make our calcu-

---

* Stepping back, you may hear a taunting ring of tautology in all this: why do things tend to average out (or is it we who, defensively, do the averaging)? If we indeed live at the limit, why should the outlandish not be the rule rather than the exception—or would that make it the (absurdist) rule? What compels throws of two fair dice to distribute themselves better and better around the mean of 7—are these averages more than descriptive: predictive, or even prescriptive? Would it not be too curious were there no concord between the nature of ordering mind and nature itself, shown by such pointings as these? Or is asking such questions hopelessly wrong-headed, putting a cartload of results in front of the causal horse harnessed to pull them? Is there actually nowhere to step back to, from the play of phenomena? We tend to relegate these edgy concerns to the margins of our thoughts and the footnotes of our books.

lated dashes along the hypotenuse from now to then, and stride along the bell curves, crest to crest.

DISTANCE AND MYSTERY

In generalizing the Pythagorean Theorem, we altered the angle between the sides, and the shapes of the figures on them, as well as the dimension of the whole. What we haven't yet touched is the hypotenuse. What if we replaced it by a jagged line?

Do you think it would be possible, given n points in the interior of a right triangle, for a path made of straight-line segments from one end of the hypotenuse through them (in some order) to the other end, to have the sum of the squares on its segments *less* than the square on the hypotenuse?

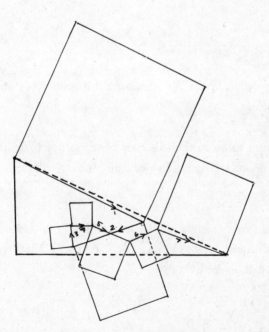

This doesn't seem very likely, at first sight: it looks as if the sum of their areas might (at least for *some* distribution of points) always mount up to be quite large indeed, like sail crowded on in a windjammer, the masts layered with canvas. Yet the statement is in fact true—and its truth is a comment on the limitations of our intuition for two dimensions. Here is the proof[3]:

Let's call a path *polygonal* if it connects a finite number of points in succession by straight-line segments. Our points will start at one end, A, of the hypotenuse, and end at the other, B, with t_1, t_2 and so on, up to t_n, in between.

If we call the path P, let $k(P)$ be the sum of the squares on its segments.

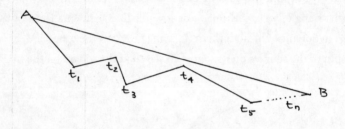

DEF.: A polygonal path P is *admissible* if it connects A and B, and if $k(P) < (AB)^2$.

We want to prove the existence of admissible paths that have at least three points on them. Divided, this problem will fall.

LEMMA: In △ABC, right angled at C, with CD the altitude to AB, if there are admissible paths H in △ADC and J in △CDB, then there is an admissible path in △ABC.

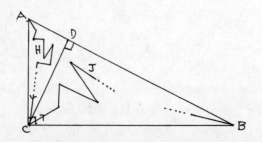

PROOF:

By admissibility, $k(H) < (AC)^2$ and $k(J) < (BC)^2$.

If HJ is path H followed by path J,

$k(HJ) < (AC)^2 + (BC)^2 = (AB)^2$, by the Pythagorean Theorem.

Yet this path isn't admissible, because C isn't an interior point of $\triangle ABC$.

However, if we remove the last segment of H (connected to C), and the first segment of J (from C), by then bridging the gap so formed with one segment from H's new end to J's new beginning, we obtain a path with yet less area of the squares on its segments:

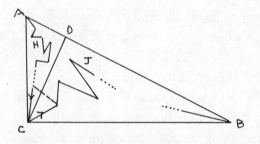

for the angle between these two segments at C can't be obtuse (since both lie within the right angle at C), so by the law of cosines the area on the new segment will be less than the sum of the two removed.

This new path (HJ, fastened together as described) is the desired admissible path for $\triangle ABC$.

We can now prove our theorem, by induction on the number of points on the path.

THEOREM: In any right triangle ABC, and for any finite number $n > 0$ of points in its interior, there is an admissible path.

PROOF:

1. If the points on our path are A, B, and Z, where Z is an interior point of △ABC, ∠AZB is obtuse, so by the law of cosines,

$$(AZ)^2 + (ZB)^2 < (AB)^2.$$

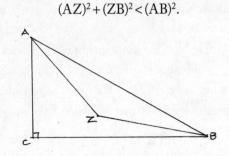

2. If there are more than two interior points, drop the altitude CD from C. Then either these points lie in both △ACD and △DCB, or not.

CASE I: If in both, then by the induction hypothesis there are admissible paths in both, and the result follows from the lemma.

CASE II: If all the interior points lie wholly in one of those two triangles—say △ACD—then construct the altitude from D to

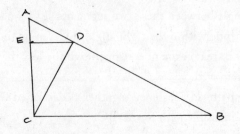

AC at E, and repeat this process if necessary, until Case I arises—as eventually it must, when a new pair of the ever smaller triangles will have hypotenuses (their maximum lengths) less than the shortest distance among the interior points of this new pair of triangles: for then at least two are guaranteed to lie in different triangles, and Case I applies.

Q. E. D.

△ △ △

What if we now let lightning strike the hypotenuse, in a storm to end all storms, so that it becomes an infinitely long lightning bolt itself. Could it possibly happen, in this apocalypse, that the sum of the areas on its segments nevertheless dwindled toward zero? This sounds even more preposterous than what just turned out not to be. How could length increase without end, like a fractal coastline, yet the area built on it amount to nothing? But mathematics is made of the unexpected, and there is a glimmer of hope in the fact that area is a square function, and if a square's side s is less than 1, then $s^2 < s$.

So let's make our zigzag path from segments of length 1, ½, ⅓, ¼, and so on forever, with the path anchored at vertex A and twisting toward B.

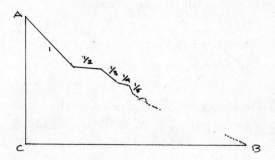

On the one hand, we know that the length of this path is infinite—the *harmonic series* $1 + ½ + ⅓ + ¼ + \dots$ diverges (to convince yourself, notice that the third and fourth terms add up to more than ½, as do the next four terms and the next eight terms after that, then the next sixteen—and so on: an endless number of halves to go along with the $1 + ½$ at the beginning).

On the other hand, these would be the areas of the squares built on those segments:

$$1 + \left(\frac{1}{2}\right)^2 + \left(\frac{1}{3}\right)^2 + \left(\frac{1}{4}\right)^2 + \dots$$

Euler was the first to discover that this infinite series converged, and to $\frac{\pi^2}{6}$ (roughly 1.64493). So if you want the sum of these areas to be less than some number, however small—say $1/n$—just subtract as many terms at the beginning as you need—1, ½, ⅓, etc.—to bring $\frac{\pi^2}{6} - \left(1 + \left(\frac{1}{2}\right)^2 + \left(\frac{1}{3}\right)^2 \text{ and so on}\right)$ under $1/n$. Then begin your mazy path at A with the first of the terms remaining. This truncated harmonic series will still diverge (you'll still be adding up an infinite number of halves), so the path will be infinitely long—but the sum of the areas on its segments will be less than the $1/n$ you required.

If you think of mathematics as housed in a gleaming cube far away, where a grid city rises on a pale horizon, don't such wares as these suggest rather that its awnings spread in a murmurous souq, or in the compiled jumble of gables and chimneys where

> Alleys John Winthrop's cattle planned
> Meander like an ampersand[4]

or in Lusitania, a shadowy Nowhere of catacombs and caverns, where custodians of truths and secrets keep the torches lit?[5] Such was the name that the followers of Nikolai Nikolaevitch Luzin, whom you met in Chapter Three, gave to their society—which brings us to . . .

△ △ △

. . . the Staircase. Earlier in this chapter, you recall, we better and better approximated a straight line by an ever more rapidly vibrating sine wave. Couldn't we modify that approach here by taking a flight of stairs instead, developed from the original triangle's tread and riser, and replace the hypotenuse by it?

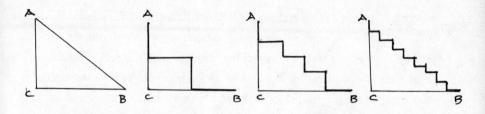

That is, make a sequence of functions of ever shorter verticals and horizontals, starting at A and going to B, and clinging more and more closely to the hypotenuse. This lets us fantasize that doing so infinitely will give us the hypotenuse itself, as the limit of these staircases, which are diminishing past shark teeth to sawteeth, and beyond. Didn't we after all do something like this before, with our little triangles, in finding the length of a curve?

What an appealing fantasy—turned appalling. For were this possible, we might as well give up mathematics on the spot as hopelessly paradoxical: those staircases—no matter how short and numerous their treads and risers—would always have a total length equal to the original triangle's two sides, while at the same time being shorter than them (by the fundamental triangle inequality).*

Luzin was plagued, in his student days, by this staircase—as many

---

* Yet might this not be just a charming flaw in an otherwise resplendent fabric—a 'spirit gate', those errors some American Indians were said to have woven deliberately into their work? No: mathematics suffers from the "dram of eale" syndrome, so devastatingly clarified by Hamlet:

> These men
> Carrying I say the stamp of one defect
>   . . . be they as pure as grace,
> As infinite as man may undergo,
> Shall in the general censure take corruption
> From that particular fault. The dram of eale
> Doth all the noble substance of a doubt
> To his own scandal.

Perhaps we can weave one true theorem after another into the tapestry of mathematics without ever completing it; but a single contradiction entangled in its tight threads unravels the whole.

another has been, before and since. "Boleslav Kornelievich!" he protested to his professor, "instead of a finite set I take a denumerably infinite set, for example, all points on the diagonal that are a rational distance from the beginning of the diagonal. Next I take all straight lines passing through the points of the set parallel to the X- and Y-axes respectively, and finally, I eliminate all 'superfluous' parts of the straight lines. What will I obtain? A saw, but one with actually infinitely small teeth! In other words, there is an individual, fixed curve that is *infinitesimally different from the diagonal.*"[6]

"Listen—what nonsense you are dishing out!" Professor Mlodzeevskii explained, and strode away.

Where did this leave Luzin—where does it leave us? Luzin had disturbed a kraken in the ocean trenches of mathematics, which is now swimming toward us with its lurid mouth agape. Around it are monsters with teeth on their teeth and yet finer teeth on those, ad infinitum, their series converging toward a function continuous everywhere and differentiable nowhere: all teeth.* Luzin's creature may not be as definitively devastating—but it is menacing enough, since its irresolvable paradox threatens to devour mathematics.

Why should sine waves of higher and higher frequency converge to a straight line, and not these ever more rapidly vibrating square waves? What you may think a negligible difference between them plays an important role: the sine waves are smooth (differentiable), and length—as you saw—is defined in terms of the derivative. If a sequence of functions, which converges point by point to a limit, has its *derivatives* converging uniformly to that limit function's derivative too, then the lengths of the graphs of the functions in the sequence will converge to the length of the limit's graph.

---

* You can follow a brilliant exposition of this story in Michael Spivak's *Calculus*, 2nd ed. (Berkeley, Calif.: Publish or Perish Press, 1980), 465–76. It is an exposition and an explanation that would have angered Luzin, since it is steeped in the juices of continuity in general and Weierstrass's approach to it in particular—and these, for Luzin, were of the devil, closing as they did those gaps of discontinuity through which such causeless darts as Grace might penetrate the world.

Unfortunately this *sufficient* condition isn't also *necessary*: we can have a sequence of functions *not* everywhere differentiable converging to a smooth limit, with its lengths converging to that limit's length. So Archimedes famously made better and better approximations to the circumference of a circle by a sequence of inscribed polygons, whose sides shortened as they grew in number. Adjacent sides did indeed meet at sharp corners that therefore had no derivative at them, but the angles at those corners grew flatter and flatter, unlike Luzin's constant right angles.

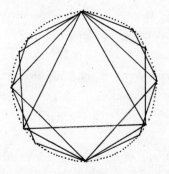

Our difficulty is growing increasingly angular itself. To smooth it away, keep before your mind's eye the image of a saw laid across a knife: no matter how many minute teeth the one may have, it isn't the other; they cut according to different principles. The functions in a convergent sequence will have many properties, not all—or even most—of which their limiting function may share (you can divide by each of the fractions $1/n$ that converge to 0 as n increases, but you can't divide by 0). Here, picture the staircase fitted within a rubber band, which tightens and narrows as the steps diminish in size. This elastic boundary

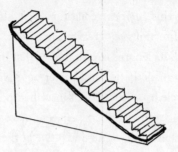

encloses the hypotenuse too, of course, and shrinks down to it, compressing the staircase as it does so—but that doesn't mean every property of the staircase—such as its length—persists with it. Haven't you just seen, in fact, an example of length going haywire as a coastline crinkles?

This cages Luzin's monster, making mathematics safe again—at least until the next kraken wakes.

△ △ △

Esprit d'escalier. A friend of ours, both a carpenter and a mathematician, told us that he had been faced with the problem of measuring a banister's length for a spiral staircase he was building: a formidable-looking task, studying the drawing—what was the formula for the length of such an arc, and what arc, after all, among all those that cycle through space, was it?

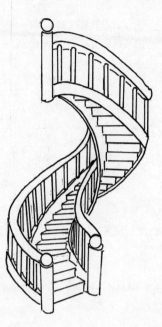

And then the mathematician in him spoke to the carpenter: slice this cylinder open:

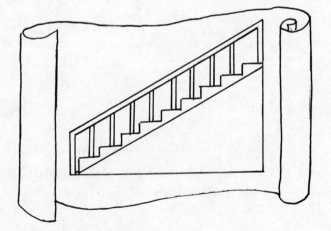

APPENDIX

(A) How $X \perp Y \Leftrightarrow \langle X, Y \rangle = 0$ follows from the law of cosines:

You saw in Chapter Six that

$$c^2 = a^2 + b^2 - 2ab \cdot \cos \gamma.$$

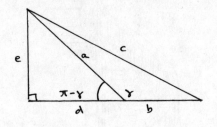

Extract from this diagram the triangle in question,

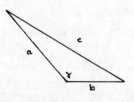

then rethink and relabel it in terms of vectors:

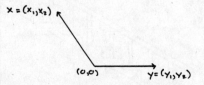

In this new language, what is the length of the old side c? Complete the parallelogram and notice that its fourth vertex is $(x_1 - y_1, x_2 - y_2)$

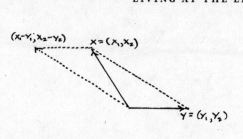

so that the vector to it is $X - Y$.

Clearly the side we want has the same length $|X - Y|$ as this, so that the law of cosines, translated into vector language, reads:

$$|X - Y|^2 = |X|^2 + |Y|^2 - 2|X| \cdot |Y| \cos \gamma.$$

Writing all but the last term out in terms of the coordinates:

$$(x_1 - y_1)^2 + (x_2 - y_2)^2 = x_1^2 + x_2^2 + y_1^2 + y_2^2 - 2|X| \cdot |Y| \cos \gamma.$$

Expanding,

$$(x_1^2 - 2x_1 y_1 + y_1^2) + (x_2^2 - 2x_2 y_2 + y_2^2)$$
$$= (x_1^2 + x_2^2) + (y_1^2 + y_2^2) - 2|X| \cdot |Y| \cos \gamma.$$

So, canceling and simplifying,

$$-2(x_1 y_1 + x_2 y_2) = -2|X| \cdot |Y| \cos \gamma,$$

so that

$$x_1 y_1 + x_2 y_2 = |X| \cdot |Y| \cos \gamma.$$

Now just dress this up in our vector finery, with $\langle X, Y \rangle$ for the inner product, and we have

$$\langle X, Y \rangle = \|X\| \cdot \|Y\| \cos \gamma.$$

Looking at the two vectors X and Y in n-space, you'd see

$$\sum_{i=1}^{n} x_i y_i = \langle X, Y \rangle = \|X\| \cdot \|Y\| \cos \gamma.$$

And if $X \perp Y$, then $\cos \gamma = 0$, so

$$X \perp Y \Rightarrow \langle X, Y \rangle = \|X\| \cdot \|Y\| \cdot 0 = 0.$$

The converse clearly holds, if neither X nor Y is the 0 vector.

(B) The inner product for Fourier series.

The inner product we had, $\langle X, Y \rangle$, was a sum of the products of these vectors' respective coordinates. Now it will be a sum of the multiplied products of f and g on the chosen interval $[a, b]$—where the analogue of sum is the integral from a to b (its symbol, after all—the elongated eighteenth-century S, $\int$—was meant to remind readers of this analogy). So

$$\langle f, g \rangle = \int_a^b f(x) g(x) dx.$$

The norm drawn from this inner product is (just like the norm on p. 198)

$$\|f\| = \sqrt{\int_a^b f(x)^2 \, dx}.$$

A natural interval over which to look at trigonometric functions is $[-\pi, \pi]$, and it turns out that for any m and n, both

$$\frac{1}{\pi} \int_{-\pi}^{\pi} \sin mx \cdot \sin nx \, dx \text{ and } \frac{1}{\pi} \int_{-\pi}^{\pi} \cos mx \cdot \cos nx \, dx$$

are 0 if $m \neq n$ and 1 if $m = n$, while $\frac{1}{\pi} \int_{-\pi}^{\pi} \sin mx \cdot \cos nx \, dx = 0$.

This tailor has killed two at a blow: establishing the perpendicularity of the axes (since $\langle \sin mx, \cos nx \rangle = 0$), and that each is of unit length. These are the $e_1, e_2, \ldots$ we had hoped for.

New technology, however, inevitably raises new technological problems, which new techniques will solve. The tight little helices of infinite Fourier series taper down (like waterspouts tilted this way and that in a da Vinci notebook) to their limits—but what assurance have we that the limits themselves will also lie in this infinite dimensional space? Might not the tips of these screws just fail to engage in the vectors they tunnel toward?

> Oh, the little more, and how much it is!
> And the little less, and what worlds away![7]

Fortunately the way our inner product behaves kills this problem dead too: how these series converge, and subtle inequalities applied to them, show that their limits are also in the space, which is therefore called *complete*: a Hilbert space (after its twentieth-century inventor). We have carried the natural bent of our thought, via induction, to its limit: at the edge, perhaps, of understanding, but within reason.

CHAPTER NINE

The Deep Point of the Dream

Every dream, Freud once wrote, has a deep point beyond which analysis cannot probe. Whether or not this is true of dreams, there are cyclones in mathematics that funnel down and down even below the ground level of apt connections and on past their intertwining roots.

We've watched the Pythagorean Theorem take shape in ways that were often starkly different from what reason would have predicted, then acquire a variety of proofs from as great a variety of people, and afterwards spread in the most far-reaching and unexpected directions. All along, however, while this theorem may have seemed an increasingly significant ganglion in Euclidean geometry, nothing suggested that it *was* Euclidean geometry itself: the geometry of flat planes, characterized by the Parallel Postulate. Now we have come to the deep point of this dream: *PT, the Pythagorean Theorem, is equivalent to the Parallel Postulate, PP!* Each, that is, implies the other, so that either could be taken as the Ancient of Days in this world of congenial insights and crisp proofs.

Hard to believe? As you saw in Chapter Five, when the forty-year-old Thomas Hobbes first came on the Pythagorean Theorem in a book lying open at a friend's house,

> "By God!", said he, "this is impossible!" So he read the demonstration of it, which referred him back to such a proposition; which proposition he read. That referred him back to another, which he also read. And so in order that at last he was demonstratively convinced of that truth.[1]

Let's follow Hobbes's example and demonstratively convince ourselves of its equivalence to the Parallel Postulate. You probably expect that a revelation this profound must be so refined or abstruse as to be all but ungraspable. In fact the story will fall out in just eight short encounters. And you'd certainly think that so deep a result, showing that what characterizes Euclidean geometry is as much the PT as the PP, would be the end if not the beginning of any introductory plane geometry course. Yet it is hardly known, and while it is evident that $PP \Rightarrow PT$ (look at any proof of the latter and you'll find the PP used somewhere along the way—as in Euclid's VI.31, where the key use of similar triangles depends on it), proofs that complete the equivalence by showing $PT \Rightarrow PP$ don't exactly abound. You can hear a few faint whispers in stray late twentieth-century journals ("That Mr. Pythagoras living down at the end of the street? Well, they say that he . . .").[2]

All but one of the steps in the proof we'll follow are due to the renowned early nineteenth-century French mathematician Adrien-Marie Legendre (who, said his colleague Poisson, "has often expressed the desire that, in speaking of him, it would only be the matter of his works, which are, in fact, his life"—he was born rich, died poor, made outstanding contributions to algebra, statistics, and number theory, and left a still unproven conjecture behind: that there is a prime number between n^2 and $(n+1)^2$, for every positive integer n). One step is the work of a Dr. Brodie.[3]

We'll sketch the sequence of proofs, then outline the strategy of each, and finish with tactical details. The postulates of Euclidean geometry are in the appendix to this chapter. You may want to check that the PP hasn't been used covertly in any of the steps along the way—all the other postulates are available.

This brings up a subtle issue. The postulates without the PP constitute what has come to be called 'neutral geometry': a geometry where through a point not on a line there is at least one parallel to that line. *Exactly one* gives us Euclid's flat plane—but more than one would be possible (hence violating the PP) on a surface with negative curvature, like a whirlpool ('hyperbolic geometry'). If, however, you think of a surface with positive curvature, such as a sphere, and on it take great circles for lines (we'll soon justify this arbitrary-seeming move), then again the PP would be false, but this time because there would be *no* parallels to a line through a point not on it. Here the postulates of neutral geometry fail: specifically, the four betweenness postulates, which have to be replaced by seven 'separation' postulates; and the notions of segment and triangle have to be redefined. These modifications constitute another non-Euclidean geometry, called *elliptic*. So what we'll now prove is that when the PT is added to the rest of the neutral geometry's postulates, the curvature of the surface becomes zero, and we find ourselves returned to Euclid's flat plane.

You see, then, that what we're about to launch into is the one sort of generalization we hadn't yet touched on: varying the kind of surface— flat or positively or negatively curved—on which our right triangle might lie. Once we complete the proof that zero curvature depends on the PP or the PT (by showing their equivalence), we will be able to look at what can be no more than *analogues* to the PT on a sphere and on a hyperbolic (negatively curving) surface. The proof has eight pieces (proposition I.16 of Euclid; four lemmas, numbered 1 through 4; three theorems, A—whose premise is the PT—B, and C, whose conclusion is the PP), and the proofs linking them together into the overall deduction: PT $\Rightarrow$ PP. The chain looks like this:

$$A \Rightarrow B \Rightarrow C$$
$$\Uparrow \quad \Uparrow$$
$$3 \quad 4$$
$$\nwarrow \quad \nearrow$$
$$I.16 \Rightarrow 1 \Rightarrow 2$$

We'll begin with a proof of I.16:

I.16 (EUCLID): In any triangle, the exterior angle is greater than either of the interior and opposite angles.

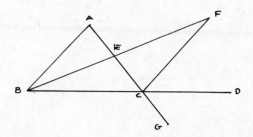

So here, we want to prove that $\angle ACD > \angle CBA$, and $> \angle BAC$.

STRATEGY: Show that each of the angles in question is congruent to another angle, which lies wholly within the exterior angle.

PROOF:

1. Bisect AC at E (so $AE \cong EC$).
2. Extend BE its own length to F (so $BE \cong EF$).
3. Construct CF.
4. Extend BC to D, AC to G.
5. $\angle AEB \cong \angle CEF$.
6. $\triangle AEB \cong \triangle CEF$.
7. $AB \cong FC$, and the remaining angles are correspondingly congruent.
8. In particular, $\angle BAE \cong \angle ECF$.
9. But $\angle ECD > \angle ECF$.
10. Hence $\angle ACD > \angle BAE \cong \angle BAC$.
11. To show $\angle ACD > \angle CBA$, repeat this process, this time bisecting BC, which (by congruent triangles) will have $\angle CBA$ congruent to an angle in the interior of $\angle BCG$, which is itself congruent to $\angle ACD$.

LEMMA I (LEGENDRE): In any triangle ABC, the sum of any two angles must be $< 180°$.

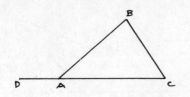

STRATEGY: Apply Euclid I.16.

PROOF:

1. Extend CA to some point D.
2. $\angle DAB + \angle BAC = 180°$.
3. But $\angle ABC < \angle DAB$ (by Euclid I.16),
4. So $\angle BAC + \angle ABC < 180°$.

LEMMA 2 (LEGENDRE): The sum of the angles in any triangle ABC is $\leq 180°$.

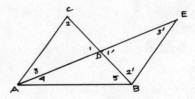

STRATEGY: This pivotal lemma is proved by assuming it false, then chasing around some angles and repeating the process until Lemma 1 is contradicted.

PROOF:

1. Assume not. Then $\angle ABC + \angle BCA + \angle CAB = 180° + a$ $(a > 0)$.
2. Assume further that $\angle ABC \leq \angle BCA$.

3. Bisect CB at D (so CD ≅ DB), and extend AD its own length to E (so AD ≅ DE).

4. ΔCAD ≅ ΔBED.

5. Numbering the angles as shown, $2 \cong 2'$, $3 \cong 3'$, and $2+3+4+5 = 2'+3'+4+5 \Rightarrow$ the sum of the angles of ΔABC = the sum of the angles of ΔABE.

6. Since we're supposing that $5 \leq 2$, $AC \leq AB$ (by Euclid I.19, the greater side subtends the greater angle), and since $AC \cong EB$ (from step 4), $EB \leq AB$ (using step 6), so $4 \leq 3'$ (by Euclid I.18, the greater angle is subtended by the greater side).

7. Hence $4 \leq 3$, and twice $4 \leq 3 + 4 \Rightarrow 4 \leq \frac{1}{2}(3+4) = \frac{1}{2}\angle CAB$.

8. We thus have a triangle, ΔABE, whose angle sum is that of ΔABC, but with one angle at most half of one of the original triangle's angles.

9. Repeat this process (bisecting the side again and again), each time getting a new triangle with angle sum equal to that of ΔABC, but with the same angle continually diminishing. Do this as many times as is necessary for that angle to be less than a.

10. In this triangle, the sum of the remaining two angles must therefore be $> 180°$, which contradicts Lemma 1, thus proving our lemma.

LEMMA 3 (LEGENDRE): If the angle sum of a triangle, ΔABC, is 180°, and a line from some vertex (say A) is drawn to meet the opposite side at D, then the angle sum of each of the two new, smaller triangles (ΔBAD, ΔDAC) is also 180°.

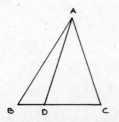

STRATEGY: Did the assumption and construction not force the conclusion, Lemma 2 would be contradicted.

PROOF:

1. Let the angles of $\triangle BAD$ sum to S_1, the angles of $\triangle DAC$ to S_2.
2. Then since $\angle BDA + \angle CDA = 180°$, the angle sum of $\triangle ABC = S_1 + S_2 - 180°$.
3. Since we assume the angle sum of $\triangle ABC = 180°$,
   $$S_1 + S_2 - 180° = 180° \Rightarrow S_1 + S_2 = 360°.$$
4. Hence if $S_1 < 180°$, we must have $S_2 > 180°$, contradicting Lemma 2.
5. Hence $S_1 = 180°$, and the same argument gives us $S_2 = 180°$.

LEMMA 4 (LEGENDRE): Given a line l and a point P not on it, there is a line m through P, intersecting l, so that the base angle formed is less than any given positive magnitude ε.

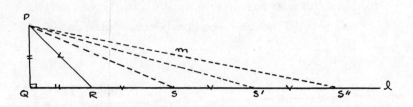

STRATEGY: Dropping a vertical from P to l, successive 'hypotenuses' from P to l make successively shallower angles with l.

PROOF:

1. Construct $PQ \perp$ to l at Q.
2. Choose R on l so that $QR \cong PQ$, and construct PR.

3. Choose S on l so that Q*R*S (* is the symbol for 'betweenness'—see the betweenness postulates in the appendix), and PR ≅ RS.

4. $\angle QRP + \angle PRS = 180° \Rightarrow \angle QRP = 180° - \angle PRS$.

5. $\angle PRS + \angle RPS + \angle RSP \leq 180°$ (Lemma 2),

6. Hence $\angle RPS + \angle RSP \leq 180° - \angle PRS = \angle QRP$ (by step 4).

7. ΔRPS is isosceles (by step 3) $\Rightarrow \angle RPS = \angle RSP$.

8. Replacing $\angle RPS$ in step 6 by $\angle RSP$, $2\angle RSP \leq 180° - \angle PRS = \angle QRP$, so $\angle RSP \leq \frac{1}{2}\angle QRP$.

9. In this way, we keep getting smaller and smaller angles where the new longer and longer 'hypotenuses' meet l, so that eventually this angle will, as desired, be less than ε, between l and the desired m.

You may feel as if you've been holding your breath while building a miniature bridge out of toothpicks. You will soon be rewarded by seeing how sturdy the finished structure is.

THEOREM A (BRODIE): PT $\Rightarrow$ The angle sum of every isosceles right triangle is 180°.

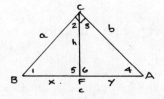

STRATEGY: Drop the altitude from the right angle, thus making two right triangles. Repeated use of the PT (making the proof long but not complicated) shows these are isosceles too. Equality of angles, and the initial right angle, give the desired result.

PROOF:

1. In right triangle ABC, right-angled at C, drop the altitude to BA, meeting it at F, and label the lengths and angles as shown.

2. Since $\triangle BCA$, $\triangle BFC$, and $\triangle CFA$ are right triangles, we have, by the PT:

$$a^2 + b^2 = c^2 \tag{1}$$
$$x^2 + h^2 = a^2 \tag{2}$$
$$h^2 + y^2 = b^2 \tag{3}$$

3. Now $c^2 = (x+y)^2 = x^2 + y^2 + 2xy$ (4)

4. So $\quad a^2 \quad + \quad b^2 \quad = \quad c^2$

$$\uparrow(2) \qquad \uparrow(3) \qquad \uparrow(4)$$

$$x^2 + h^2 \quad + \quad h^2 + y^2 \quad = \quad x^2 + y^2 + 2xy$$

5. So $2h^2 = 2xy \Rightarrow h^2 = xy \Rightarrow \dfrac{h}{x} = \dfrac{y}{h}$, and call this ratio k.

6. Hence $h = kx$ and $y = kh$, and

7. $b^2 = h^2 + y^2 = k^2 x^2 + k^2 h^2 = k^2(x^2 + h^2) = k^2 a^2$, so $b = ka$.

$$\uparrow \qquad \uparrow \qquad\qquad\qquad \uparrow$$
$$(3) \qquad \text{step 6} \qquad\qquad (2)$$

8. That means $\dfrac{b}{a} = k = \dfrac{h}{x} = \dfrac{y}{h}$.

$$\uparrow\uparrow \qquad \uparrow$$
$$\text{step 7 \quad step 6}$$

9. Similarly, $\dfrac{b}{c} = \dfrac{h}{a} = \dfrac{y}{b}$.

10. Now since our triangle is isosceles, $a \cong b$, hence $\angle 1 \cong \angle 4$.

11. And by step 8, $\dfrac{a}{b} = \dfrac{x}{h}$, and since $\dfrac{a}{b} = 1$, $x = h$. Similarly, $h = y$.

12. Thus $\triangle BFC$ and $\triangle AFC$ are also isosceles.

13. Hence $\angle 1 \cong \angle 2 \cong \angle 3 \cong \angle 4$.

14. $\angle 5 = \angle 6 = 90°$ (by construction),

15. Hence $\triangle BFC$, $\triangle CFA$ and $\triangle BCA$ are equiangular with each other.

16. But $\angle 2 + \angle 3 = 90°$, so since $\angle 2 \cong \angle 1$ and $\angle 3 \cong \angle 4$, $\angle 1 + \angle 4 = 90°$.

17. Hence the angle sum of right $\triangle ABC$ is

$$\angle 1 + \angle 2 + \angle 3 + \angle 4 = 180°.$$

THEOREM B (LEGENDRE): If the angle sum in any isosceles right triangle is 180°, then so is the angle sum in any right triangle.

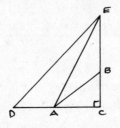

STRATEGY: Constructing an isosceles right triangle containing the given right triangle, followed by one application of Theorem A and two of Lemma 3, yields our result.

PROOF:

1. Extend CA to D and CB to E so that CE ≅ CD; hence ΔDCE is isosceles and (by Theorem A) its angle sum is 180°.
2. Construct EA.
3. By Lemma 3, the angle sum of ΔEAC is 180°.
4. By Lemma 3 again, the angle sum of ΔACB is 180°.

THEOREM C (LEGENDRE): The angle sum of any right triangle is 180° ⇒ PP.

STRATEGY: A proof by contradiction, via Lemma 1 to get one parallel to *l* through P, no others via Lemma 4 (relying heavily on betweenness properties).

PROOF:

1. Construct PQ ⊥ to *l* at Q.
2. Construct line *m* ⊥ to PQ at P.

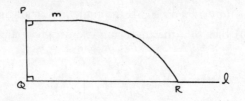

3. $m \parallel l$: for if not, m intersects l at some point R, and $\triangle$PRQ contains two right angles, contradicting Lemma 1.

4. Given steps 1 and 2, if t is another line through P, with $t \parallel l$, then choose R on t so that $\angle$RPQ $< 90°$ (for if all choices of

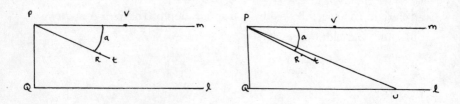

   R made $\angle$RPQ $= 90°$, $t = m$, contradicting the choice of t); and let V $\neq$ P be any point on m, on the same side of PQ as R.

5. Let a be the magnitude of $\angle$RPV.

6. By Lemma 4, there is a line through P, intersecting l (at U) such that $\angle$QUP $< a$.

7. Since $\angle$PQU is a right angle, by our hypothesis (which is Theorem B), $\angle$QUP $+ \angle$QPU is a right angle.

8. But $\angle$QPV is also a right angle, and $\angle$QPV $= \angle$QPU $+ \angle$UPV.

9. Hence (by steps 7 and 8) $\angle$QUP $\cong \angle$UPV.

10. So (by step 6) $\angle$UPV $< a$.

11. Since $\angle$RPV $= a$, the ray PR must lie in the interior of $\angle$QPU ("lines lying in the interior of angles" is another one of those betweenness issues, whose proofs are little square dances called to the tune of the four betweenness postulates).

12. Hence PR must intersect l—in fact, somewhere between Q and U (which follows from the "Crossbar Theorem": see this chapter's appendix for a proof).

13. This contradicts the supposition that $t \parallel l$, so the parallel m to l through P must be unique, which is the PP.

$$\triangle \quad \triangle \quad \triangle$$

Oh you enthusiasts of the Flat Earth Society—if any of you are still contriving with us—what a marvelous undertaking was yours! How subtle your reasoning, how intricate your works, how far—even to infinity—your conclusions stretched! The lively culture of Euclidean geometry endorses your dedication, and you have here seen your faith strengthened by adding the authority of Pythagoras to it. Even more glorious, however, is the civilization of which that culture is a part: for as with everything in mathematics, what was taken to be the whole inevitably comes to be seen as involved in a greater, filled with similar, then ever more diverse cultures, evolving together harmoniously toward comprehension: Mind thinks One.

Here then is a universe parallel to Euclid's: a sphere. What creatures aping the Pythagorean inhabit it? It models the postulates for the 'elliptic geometry' we spoke of before, if we interpret lines as great circles (circles on the sphere's surface, that is, centered at the sphere's center, O), and agree to think of two antipodal points as the same (pasting them together makes these postulates work, although visualizing this is impossible for our Euclidean eyes).*

---

* What right have we to interpret lines on the sphere as great circles, or to say that their two intersections are just one? What right have the French to call bread *pain*? That's their interpretation of an object in the world, and it seems to work as consistently in their language as 'bread' does in ours. French and English are two incarnations of such axioms as language has. These axioms, like those for spherical geometry, don't come with an ordained instantiation. Anything that faithfully represents them will do. You saw in Chapter Eight some of the widely different ways of interpreting the postulates for measuring distance. Here the choices for picturing lines on a sphere are more constrained, but what's pleasing is that *interpreting* them as great circles neatly satisfies all the postulates of elliptic geometry without unnecessary distortions. After all, we could have interpreted lines on the Euclidean plane as cascades of ripples, or as solution sets to the equations $y = mx + b$,

On the sphere there will be no lines parallel to a given line (great circle) through a point not on it—any two great circles intersect ("once"—in those two antipodal points). The angle-sum of a triangle, as you might guess, won't be less but more than 180°.

So here is a spherical triangle, CBA—think of those vertices as the

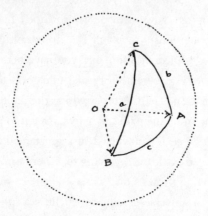

arrow-ends of vectors from the sphere's center, O. We'll call the triangle's sides a, b, and c, as shown, with $\angle C$ opposite side c. We're after the analogue of the Pythagorean Theorem—a relation among the sides a, b, and c, when C is a right angle—and all our previous work with vectors will help us now to find the 'spherical law of cosines', which is the tool for this job.

Finding, polishing, and using this tool is in the workshop of the appendix. What it will uncover is this remote likeness of the theorem we have come to know so well: on a sphere of radius 1, for a triangle with sides a, b, and c, right-angled at C,

$$\cos c = \cos a \cdot \cos b.$$

but the ornateness of the one, the abstraction of the other, might have blunted our intuition. The axioms are like regulative thoughts, expressed time and again in shapes as various as that of the hero Euphorbus and the herald Aethalides.

Should you want the analogue of the PT on a sphere of radius R, then the radian measures of the spherical triangle's sides are no longer a, b, and c but $\dfrac{a}{R}$, $\dfrac{b}{R}$, and $\dfrac{c}{R}$, and the theorem becomes

$$\cos\left(\frac{c}{R}\right) = \cos\left(\frac{a}{R}\right) \cdot \cos\left(\frac{b}{R}\right).$$

You may walk away disappointed after so much work, thinking that this looks very little like the Pythagorean Theorem. Well, we're no longer on the plane—or are we? Here is the first of two great surprises.

What if we let the sphere's radius R increase, so that its surface becomes ever less curved? Recall (from Chapter Five) the Taylor series for cosine:

$$\cos x = 1 - \frac{x^2}{2!} + \frac{x^4}{4!} - \frac{x^6}{6!} + \cdots$$

Substituting this in the equation above,

$$1 - \frac{c^2}{2R^2} + \frac{c^4}{24R^4} - \frac{c^6}{720R^6} + \cdots =$$
$$\left[1 - \frac{a^2}{2R^2} + \frac{a^4}{24R^4} - \frac{a^6}{720R^6} + \cdots\right]\left[1 - \frac{b^2}{2R^2} + \frac{b^4}{24R^4} - \frac{b^6}{720R^6} + \cdots\right]$$

Factor this (using 'terms' to stand for a sum of terms all of whose denominators contain some power of R):

$$1 - \frac{c^2}{2R^2}(1 + \text{terms}) = \left[1 - \frac{a^2}{2R^2}(1 + \text{terms})\right]\left[1 - \frac{b^2}{2R^2}(1 + \text{terms})\right].$$

Multiply out:

$$1 - \frac{c^2}{2R^2} - \text{terms} = 1 - \frac{b^2}{2R^2}(1 + \text{terms}) - \frac{a^2}{2R^2}(1 + \text{terms})$$
$$+ \frac{a^2 b^2}{4R^4}(1 + \text{terms})(1 + \text{terms}).$$

Now subtract 1 from both sides, then multiply by $2R^2$:

$$-c^2 - \text{terms} = -b^2 - \text{terms} - a^2 - \text{terms} + \frac{a^2 b^2}{2R^2} + \text{terms}.$$

Let R—the sphere's radius—go to infinity. Then since R is in the denominator of every term, and of $\frac{a^2 b^2}{2R^2}$, these denominators will each go to infinity, and so all of the terms, along with $\frac{a^2 b^2}{2R^2}$, will go to zero, and we will be left with:

$$-c^2 = -b^2 - a^2.$$

Multiply by −1:

$$c^2 = a^2 + b^2.$$

We live at the limit.

△ △ △

The second great surprise lies around a couple of corners: the first takes us to the hyperbolic plane, the second to complex space.

Planes bending away from the Euclidean—the spherical with positive, the hyperbolic with negative curvature—there must be the sweetest of symmetries here. Since we've just found that the spherical analogue of the Pythagorean Theorem is

$$\cos c = \cos a \cdot \cos b,$$

we might hope that for cosh, the hyperbolic cosine function, we would have

$$\cosh c = \cosh a \cdot \cosh b.$$

We won't be disappointed. But while the distance from here to there isn't a year and a half, there is considerable ground to cover: what in fact does the hyperbolic plane look like? How should distance be measured on it, so that '*a*', '*b*', and '*c*' make sense? And exactly what are these analogous trig functions?

To say for a start what the plane looks like will take some doing. Just as any map of the sphere onto a flat surface entails unavoidable distortion (think of Mercator's generous polar regions), so will picturing the hyperbolic plane—even in three dimensions. We have several models of it, which don't look anything like one another.* A significant distortion all share is this. Like the Euclidean plane and the sphere, the hyperbolic plane is highly homogeneous: which implies, for example, that in a right triangle anywhere on it, two of its sides will determine the third. This is, after all, the core of the Pythagorean Theorem, but you might not have guessed it was true on the hyperbolic plane by looking at any one of its incarnations, like the *pseudosphere*, where triangles down toward the bell will look so different from those farther up the tapering pipe.

To speak of triangles on the pseudosphere means understanding the lengths of their sides. Since using the Pythagorean ruler works only

---

* All are due to the nineteenth-century Italian geometer Eugenio Beltrami, although some have been named after their rediscoverers. Beltrami had variously been secretary to a railway engineer, a professor of rational mechanics, and a senator of the Kingdom of Italy (the contrast between these last two must have taken him aback).

on a flat plane, we can't expect to apply it here. Distance will still be measured along the shortest path from point to point—the *geodesic*—but on the pseudosphere this path is called a *tractrix*, and is just the pseudosphere's profile:

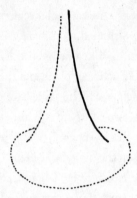

Think of it as traced by a weight starting out at a table's upper left-hand corner and tugged along it by a string hanging over the table's bottom edge, pulled slowly but steadily to the right. The sides '*a*', '*b*', and '*c*' of the inward-curving triangle on a pseudosphere are segments of such tractrices, and some ingenuity yields a metric on them (a measure satisfying those axioms for distance we talked about in Chapter Eight, p. 205–206).*

What will the trigonometric functions on a pseudosphere be? Since the standard trig functions are defined on a circle,

---

* We won't let it trouble us that triangles on the hyperbolic plane have many peculiar features, such as that their area remains finite even when all their sides are infinite (although Lewis Carroll's White Queen could believe six impossible things before breakfast, this was one that Reverend Dodgson couldn't accept at any time of day, concluding that non-Euclidean geometry was non-sense). In our single-minded quest for the Pythagorean relation, all that matters is that we've managed to cook up some way of measuring distance here, and so can speak of the side-lengths of triangles and the ratios among them.

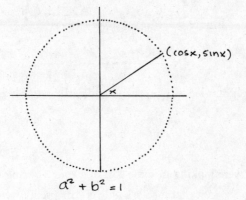

$$a^2 + b^2 = 1$$

the analogous functions here will be defined on a hyperbola:

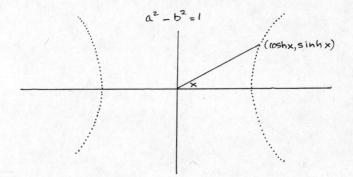

Letting 'sinh' abbreviate 'hyperbolic sine', just as 'cosh' did 'hyperbolic cosine', if we relate them to e^x (as you saw on p. 157 Euler's hybrid orchid did with sine and cosine), by cross-fertilizing their Taylor series with that for the exponential function, we will see blossom

$$\sinh x = \frac{e^x - e^{-x}}{2} \quad \cosh x = \frac{e^x + e^{-x}}{2}$$

And look what this tells us*:

---

* What follows as well from the power series for sin, cos, and e is that for *complex* numbers z, $\sin^2 z + \cos^2 z = 1$.

$$\cosh^2 x - \sinh^2 x = \frac{e^{2x} + 2e^x e^{-x} + e^{-2x} - e^{2x} + 2e^x e^{-x} - e^{-2x}}{4}$$

$$= \frac{2e^x e^{-x} + 2e^x e^{-x}}{4} = \frac{4e^0}{4} = 1.$$

Just as our diagram led us to expect—and eerily reminiscent of

$$\sin^2 x + \cos^2 x = 1.$$

It is as if we were looking in some sort of a transfiguring mirror. What we want is to see in it that for a triangle on the pseudosphere with sides a, b, and c, and right-angled at C,

$$\cosh c = \cosh a \cdot \cosh b.$$

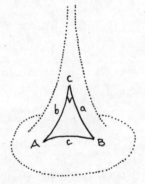

While we could certainly do this by imitating the proof we had on the sphere, it would take longer, only because the geometry on such an unusual surface as this isn't as clearly fixed in our thought as is the geometry on the Euclidean plane or the sphere.* But by looking in this mirror, we will see not only the hyperbolic analogue of the Pythagorean Theorem,

---

* The easiest of the various ways to carry this out would be on a different model of the hyperbolic plane—the so-called Poincaré disk, which is the interior of a circle C, where the 'points' are Euclidean points but 'lines' are the arcs of other circles that meet C at right angles (along with the diameters of C), not including the points on C's circumference.

but the whole of the marvelous duality between sphere and pseudosphere, elliptic and hyperbolic geometry, and the movement here from the first to the second. The whole: we won't therefore see the engineering—how the parts articulate, or even the shapes of those parts—but will catch the grand gist of this transformation. So a question answered in terms of *why* rather than *how* satisfies our need for meaning if not knowledge. Here then we round that second corner, bringing us back, as promised, to the garden of complex space in Chapter Six.

For this mirror is the imaginary numbers, whose reality—as gauged by their ubiquity, profundity, and power—is in fact greater than that of the reals they underlie.

You remember that after finding the spherical Pythagorean relation, we generalized from a unit sphere to one of any radius R—so that eventually $\dfrac{1}{R^2}$ became a factor when we multiplied everything out (p. 237). In fact, $\dfrac{1}{R^2}$ is the *constant of curvature k* of a sphere of radius R. Since the curvature of a pseudosphere is negative, its curvature constant $k' = \dfrac{-1}{R^2}$. As a magician would say, reaching into his hat, watch carefully. To go from k to k', multiply R by i:

$$k = \frac{1}{R^2} \rightarrow k' = \frac{1}{(i\mathrm{R})^2} = \frac{-1}{R^2}.$$

This transforms the sphere into the pseudosphere, and at the same time metamorphoses the spherical into the hyperbolic trig functions! How?

Take Euler's exotic orchid once again, but this time inject his formula for cosine with the complex variable z. This is what will grow:

$$\cos z = \frac{(e^{iz} + e^{-iz})}{2}.$$

And now, multiplying throughout by i, we have our transformation:

$$\cos z = \frac{(e^{iz} + e^{-iz})}{2} \rightarrow \frac{(e^z + e^{-z})}{2} = \cosh z.$$

↑

multiply variable by i

So the spherical Pythagorean relation, $\cos c = \cos a \cdot \cos b$, morphs into its hyperbolic double: $\cosh c = \cosh a \cdot \cosh b$.

The sphere, when inflated, with $R \rightarrow \infty$, approached the Euclidean plane and its Pythagorean relation, $c^2 = a^2 + b^2$. Here, as $R \rightarrow \infty$ and the hyperbolic triangles look smaller and smaller, the hyperbolic plane's ever-lessening negative curvature approaches the zero curvature of the Euclidean, whose Pythagorean relation is likewise approached by its hyperbolic analogue.

Our having stepped back has a well-known advantage for better leaping. While for spherical geometry there is only one sphere (its constant positive curvature is all that matters, not the varying radii of its models), you may rightly be puzzled by what seem hyperbolic surfaces with many different negative curvatures—why might these not affect the geometry on them? And how, after all, do cosh and sinh really relate to the underlying curvature?

Instead of puzzling that out here, we could relish our backward step as out of embodiments and into form, saying: there is only one sphere which shows itself as the hyperbolic plane; its radius is i, and the formulas on it for cosh and sinh build an analogous geometry every bit as intricate, revelatory, and consistent as that on the sphere of radius 1.

With this sign you can go forward again, and conquer.

△ △ △

It may have been true for our grandparents that Gert's poems were punk, Ep's statues were junk, and nobody understood Ein. Ninety years on, however, it would be hard to navigate the universe without grasping the theory of relativity.

Einstein suggested—and numerous readings and experiments have since confirmed—that two observers moving uniformly with respect to each other, and momentarily coinciding, won't agree on the time t or the place (x, y, z) where an event they both see occurs—but* they *will* agree on the *interval*, I, that replaces *distance*:

$$t^2 - (x^2 + y^2 + z^2) = I^2, \text{ a constant.}$$

Were it not for that minus sign, how Pythagorean their agreements would be—but the hyperbolic curvature of spacetime accounts for this difference. If we simplify for a moment to a single spatial dimension x, there's that minus sign showing up, as before, with a hyperbola in the x-t plane:

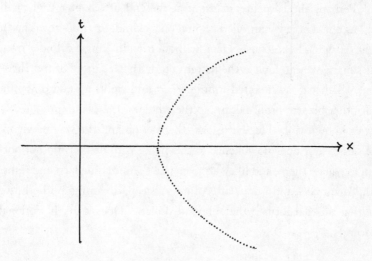

Now turn the hyperbola back into a four-dimensional hyperboloid, spun around the time axis t, and you have our universe.

We can't help being unsettled by this at first because we're used to seeing and moving in our small, flat corner of spacetime—but physics, on its armature of mathematics, urges us on to think hyperbolically, act Euclideanly.

---

* Having first chosen the same units for measuring time and space—units that conveniently make the speed of light 1.

THE RETURN OF THE VASISTDAS

What's that scrabbling behind the panels in the attic? We now know that the Pythagorean Theorem holds only on the flattest of surfaces—yet didn't we see it at the end of Chapter Five clinging to the doubled bend of a torus? Will *this*, therefore, be the moment when mathematics self-destructs?

Look back at that torus (pp. 135–37). It was made by scrolling the square on the hypotenuse into a cylinder, and then pulling that around in the other direction, joining the cylinder's top and bottom circles. Try it. What went wrong? There wasn't enough cylindrical length actually to carry out this second manoeuvre. The 1×1 proportions of the original square wouldn't allow it.

Perhaps the carpenter's insight at the end of Chapter Eight will help us out. There we unrolled a spiral on a cylinder to find it was just a right triangle's hypotenuse. Here, without even making a cylinder from our square, we can look at the pattern on it: Thabit's proof of the Theorem, which we saw iterated infinitely left and right, up and down, in that other bizarre proof taking up the whole of the plane (pp. 94–98). We've known since Heraclitus that the way up and the way down are the same, so it does no violence to this plane with neither top nor bottom to equate them, and likewise to equate their missing left- and right-hand margins—and in our after-dinner's sleep, dreaming on both, we see that this flat torus *is* the unbounded plane. Does it live in notional space?

A POINTING

Here is a curious point raised by two concerns a few pages back—or more aptly put—a pointing beyond the deep point of this dream.

We said that the hyperbolic plane has been variously modeled, with the same relations holding among parts that look so different, making the models hardly seem to be reflecting a common object.

And we passed from the sine and cosine functions to their hyperbolic equivalents via manoeuvres with complex numbers, which allowed us to write (with the complex variable z)

$$\sin^2 z + \cos^2 z = 1.$$

This was an aside we made in a footnote, even though its implications are momentous: for this equation can't represent the Pythagorean Theorem, since on the complex plane it has nothing to do with a right triangle and the areas of squares on its sides!

What, then, has it to do with? Can we say: just as hyperbolic geometry—or indeed any geometry, and any province of mathematics—is a system of relationships showing itself in this model or that, one medium (numbers, shapes) or another, so the Pythagorean *form* is an invisible attunement of parts, made perceptible (like harmony in an instrument) through different embodiments?

A consequence of this view—as perplexing to us as it was to the Pythagoreans of Plato's day[4]—is that form doesn't exist *independent* of its incarnations, nor is it somehow passed on from one to another, but (as with all things adjectival) belongs with what can be named.* We wrote in Chapter Eight of the form remaining, no matter how various the content. Remaining where? Putting the same thought more vaguely now better reflects the condition of our comprehending: for we are trying to look at what we are looking with.

APPENDIX

(A) Hilbert's Postulates for Euclidean Geometry (the first four sets constitute the postulates for Neutral Geometry).

---

* An analogy would be to song form, ABA: one melody is followed by a second, then the first repeats. This simple and universal template evolves into sonata form, where ABA has become exposition, development, recapitulation: a format within which the most far-reaching musical adventures happen. It is as if form had passed from facilitating to underlying—or is the formal cause neither efficient nor final but—formal?

INCIDENCE POSTULATES:

I-1. If P and Q are different points, then there is a unique line *l* through them.

I-2. There are at least two points on every line *l*.

I-3. There are three distinct points that don't all lie on the same line.

BETWEENNESS POSTULATES (let A*B*C stand for: "Point B is between points A and C"):

B-1. If A*B*C, then A, B, and C are distinct points all on the same line, and C*B*A.

B-2. If B and D are different points, then there are points A, C, and E lying on line BD such that A*B*D, B*C*D, and B*D*E.

B-3. If A, B, and C are different points on a line, then one and only one of the points is between the other two.

B-4. For every line *l* and any three points A, B, and C not on *l*,

 (i) if A and B are on the same side of *l*, and B and C are on the same side of *l*, then A and C are on the same side of *l*;

 (ii) if A and B are on opposite sides of *l*, and B and C are on opposite sides of *l*, then A and C are on the same side of *l*.

CONGRUENCE POSTULATES:

C-1. If A and B are distinct points and A′ is any point, then for each ray *r* emanating from A′, there is a *unique* point B′ on *r* such that B′ ≠ A′, and AB ≅ A′B′.

C-2. If AB ≅ CD and AB ≅ EF, then CD ≅ EF. Moreover, every segment is congruent to itself.

C-3. If A*B*C, A′*B′*C′, AB ≅ A′B′ and BC ≅ B′C′, then AC ≅ A′C′.

C-4. Given any angle ∠BAC, and given any ray A′B′ emanating from a point A′, then there is a unique ray A′C′ on a given side of line A′B′ such that ∠B′A′C′ ≅ ∠BAC.

C-5. If ∠A ≅ ∠B and ∠A ≅ ∠C, then ∠B ≅ ∠C. Moreover, every angle is congruent to itself.

C-6 (SAS). If two sides and the included angle of one triangle are congruent respectively to two sides and the included angle of another triangle, then the two triangles are congruent.

CONTINUITY POSTULATES:

Archimedes: If AB and CD are any segments, then there is a number n such that if segment CD is laid off n times on the ray AB emanating from A, then a point E is reached where $n \cdot CD \cong AE$ and $A*B*E$.

Dedekind: Suppose that the set of all points on a line l is the union $\Sigma_1 \cup \Sigma_2$ of two non-empty subsets such that no point of Σ_1 is between two points of Σ_2, and vice versa. Then there is a unique point O lying on l such that $P_1{}^*O^*P_2$ if and only if P_1 is in Σ_1 and P_2 is in Σ_2 (this says that if the points on a line fall into two non-overlapping sets, there will be a unique point between these two sets).

THE PARALLEL POSTULATE (Playfair's version):

Through a given point P not on a line l, only one line can be drawn parallel to l.

(B) THE "CROSSBAR" THEOREM

The Crossbar Theorem states, reasonably enough, that you can't imprison a ray from a triangle's vertex inside the triangle: it has to escape through the opposite side. That is:

If a point D is in the interior of ∠CAB, then ray AD will somewhere intersect the line segment CB.

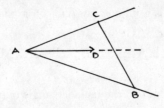

To prove this, we need a prior theorem (which follows, with a proof by contradiction, from the betweenness postulates): If D is in the interior of ∠CAB, and if C*A*E, then B is in the interior of ∠DAE.

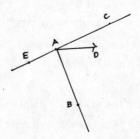

We can now prove the Crossbar Theorem by assuming it false, so that B and C would lie on the same side of AD. Now choose E on CA so

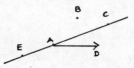

that C*A*E. By the theorem above, B is in the interior of ∠DAE, putting B and C on opposite sides of ray AD. This contradiction proves the Crossbar Theorem.

(C) THE PYTHAGOREAN THEOREM ON A SPHERE

Take O as the center of a sphere of radius 1, with △ABC, right-angled at C, on its surface. Its sides are arcs a, b, c as shown, and their lengths are the same as the measures, in radians, of the angles at O that subtend them (so arc a, for example, subtends ∠BOC).

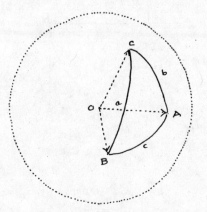

An angle, like C, in a spherical triangle is measured by the angle between the tangents to the sphere along the arcs of the sides meeting at that vertex: so here, tangents to a and b, meeting at C.

We construct those tangents as follows: in the plane OCB, draw CP⊥OC at C, meeting OB (extended) at P. Likewise, in plane OCA, draw CQ⊥OC, meeting OA (extended) at Q. CP and CQ are the tangents defining ∠C.

Let x be the length of PQ.

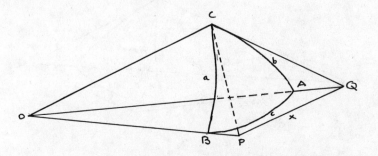

We will get the Pythagorean Theorem on a Sphere via the "Spherical Law of Cosines", which in turn will follow from Euclid's law of cosines applied to two different planar triangles in our diagram. We therefore need to determine the lengths of all their sides.

Giving a central angle the same name as the arc it subtends, we have in ΔOCP:

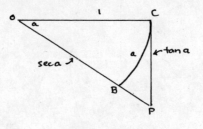

So

$$CP = \tan a$$
$$OP = \sec a.$$

Similarly,

$$CQ = \tan b$$
$$OQ = \sec b.$$

Here is our measured setup:

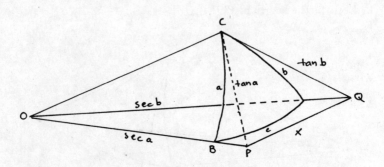

In $\triangle CPQ$, by the planar law of cosines,

$$x^2 = \tan^2 a + \tan^2 b - 2\tan a \cdot \tan b \cdot \cos C \qquad (1)$$

Similarly in $\triangle POQ$,

$$x^2 = \sec^2 a + \sec^2 b - 2\sec a \cdot \sec b \cdot \cos c \qquad (2)$$

Equating (1) and (2),

$$\sec^2 a + \sec^2 b - 2\sec a \cdot \sec b \cdot \cos c = \tan^2 a + \tan^2 b$$
$$- 2\tan a \cdot \tan b \cdot \cos \mathrm{C},$$

so

$$2\sec a \cdot \sec b \cdot \cos c = \sec^2 a + \sec^2 b - \tan^2 a - \tan^2 b$$
$$+ 2\tan a \cdot \tan b \cdot \cos \mathrm{C} \qquad (3)$$

But

$$\sec^2 a - \tan^2 a = \frac{1}{\cos^2 a} - \frac{\sin^2 a}{\cos^2 a} = \frac{(1 - \sin^2 a)}{\cos^2 a} = \frac{\cos^2 a}{\cos^2 a} = 1.$$

Likewise

$$\sec^2 b - \tan^2 b = 1,$$

so (3) becomes

$$2\sec a \cdot \sec b \cdot \cos c = 2 + 2\tan a \cdot \tan b \cdot \cos \mathrm{C}.$$

Dividing both sides by 2 and multiplying by $\cos a \cdot \cos b$, we have:

$$\cos c = \cos a \cdot \cos b + \sin a \cdot \sin b \cdot \cos \mathrm{C},$$

the Spherical Law of Cosines.
 When C is a right angle, $\frac{\pi}{2}$, $\cos c = 0$, and we have

$$\cos c = \cos a \cdot \cos b,$$

the Spherical Pythagorean Theorem.[5]

CHAPTER TEN

Magic Casements

Charmed magic casements, opening on the foam
Of perilous seas, in faery lands forlorn.
—KEATS, "ODE TO A NIGHTINGALE"

The begottens and begets of mathematics never end—not because of some dry combinatorial play, but because curiosity always seeks to justify the peculiar, and imagination to shape a deeper unity. Jefferson stood at the windows of his Palladian Monticello and looked out at the untamed frontier. These are some of the windows opening from the House of Pythagoras.

SIERPINSKI'S QUERY

In 1962 Waclaw Sierpinski, the Polish mathematician whose excessively leaky gasket you saw in Chapter Six (a patriot who had refused to pass examinations in the Russian language, worked with Luzin, taught in the Underground Warsaw University during World War II, and remained cheerful for all that half his colleagues and students had been murdered by the Nazis; he disliked any corrections to his papers) asked if there were infinitely many Pythagorean triples which were all triangular numbers. Are there *any*? Yes: we know—such is human ingenuity—that 8778, 10,296, and 13,530 are each triangular, and the sum of the squares of the first two equals the square of the third. And that's all we know.

THE PERFECT BOX

You're aware from your circle of acquaintances how different 6 or 60 or 600 people can be—or 60,000, if you look around a football stadium on a fall Saturday. But 6 million, or 6 billion? Well, all the people in our world must be different from one another, and in many ways, though it's hard to imagine the details. Harder to see how 60 billion beetles could really differ—and how, other than just in their names, the elements of yet greater collections of mere *numbers* could each have significant traits.

The story is well known about the Oxford mathematician Hardy visiting his extraordinary protégé Ramanujan in the hospital, and remarking that the number of the taxi he had taken, 1729, was rather dull. "No, Hardy," said the dying Ramanujan, "it is a very interesting number. It is the smallest number expressible as the sum of two cubes in two different ways." But what if Hardy's cab number had been 13,852,800? Or 211,773,121?

You remember the rectangular box of Chapter Six, which we traded in for all of space in Chapter Eight. Here it is again, like one of those stubborn gifts of folklore that you just can't give away. We found that we could climb up out of the plane on its body diagonal, and then up an endless Pythagorean spiral of diagonal steps into n dimensions. Since its flattened version can have all three sides be whole numbers in infinitely many ways (the Pythagorean triples), might there not be at least one way for the seven relevant lines of its three-dimensional version to be whole numbers too—the three edges a, b, and c, the three face diagonals, and the body diagonal?

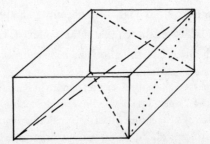

In 1719 a German accountant named Paul Halke found a box with all its edges and face diagonals whole numbers: $a = 44$, $b = 117$, and $c = 240$ (giving the face diagonals 125, 244, and 267)—but the body diagonal, $\sqrt{1936 + 57600 + 13689} \sim 270.60118$, was irrational.

The blind eighteenth-century English mathematician Nicholas Saunderson, inventor of a calculating machine called his Palpable Arithmetic, who lectured on optics, among other subjects, at Cambridge University (the man he replaced "was dismissed for having too much religion, and Saunderson preferred because he had none", wrote his friend, the astronomer Halley) came up, in the sixth book of his *Elements of Algebra*, with a way—now called Saunderson's Parametrization—for finding many rectangular boxes with whole numbers for all but the body diagonal. Halke's box is the smallest of these. Unfortunately, in 1972 W. G. Spohn, at Johns Hopkins, proved that no Saunderson box can have an integral body diagonal.

Our species has not been idle. What about boxes with only one edge irrational? In 1949 "Mahatma" (in a London journal for assistant masters) asked readers for solutions, and received an infinity of them—one of which (oh Hardy!) involves the numbers 13,852,800 and 211,773,121.

Think back to those mistaken readings of Plimpton 322 by Neugebauer, which included values for Pythagorean triples like 12,709. Why does this Theorem call up such gigantic figures from Nightmare Land? Has *every* number the mark of personality on its brow? If not each separate number, what never fails to startle is the presence among them of subtle structure: many of Mahatma's sextuples form cycles of four—and if we turn to sextuples with only a face diagonal irrational, solutions occur in cycles of five!

But a perfect box? No Mahatma nor any palpable arithmetic has yet found one, nor has yet shown that none can exist. Informed opinion has it that "for the moment, this problem still seems out of reach."*

---

* Should the madness seize you, you can read more in Richard Guy, *Unsolved Problems in Number Theory*, 3rd ed. (Berlin: Springer, 2004), section D18; and see Oliver Knill's "Treasure Hunting for Perfect Euler Bricks", http://www.math.harvard.edu/~knill/various/eulercuboid/lecture.pdf.

THE FERMAT VARIATIONS

You may have wondered why, of all the generalizations we've explored, no mention has yet been made of the most famous: changing the exponents in $x^2 + y^2 = z^2$ from 2 to higher powers. Fermat's centuries-old boast that there are no non-zero integral solutions x, y, and z for any exponent beyond $n = 2$ was answered in 1995 (how oddly dates sit among timeless truths) by Andrew Wiles and Richard Taylor, who proved that Fermat was right. Long before, Euler may have been the first to show no solutions exist in the particular case $n = 3$.

But what about possible solutions for $x^3 + y^3 = dz^3$, where d is some natural number other than 1? Or for other $n > 2$: $x^n + y^n = dz^n$, or for mixed exponents in this format? Too much of a muchness: since the case $n = 3$ has proven difficult enough (we have only the most tantalizing clues), we'll now look just at the state of play for it.

We know, for example, that of the infinitely many possible values for d, when $d = 7$, what works is $x = 2$, $y = -1$, and $z = 1$:

$$2^3 + (-1)^3 = 7 \cdot 1^3.$$

For $d = 13$ we can have $x = 7$, $y = 2$, and $z = 3$:

$$7^3 + 2^3 = 351 = 13 \cdot 3^3.$$

If $d = 7^2 = 49$,

$$11^3 + (-2)^3 = 49 \cdot 3^3,$$

and for $d = 13^2 = 169$,

That informed opinion is Robin Hartshorne and Ronald van Luijk's, in their "Non-Euclidean Pythagorean Triples, A Problem of Euler's, and Rational Points on K3 Surfaces", remark 5.4, http://www.math.leidenuniv.nl/~rvl/ps/noneuc.pdf.

$$8^3 + (-7)^3 = 169 \cdot 1^3.$$

There is a growing anthology of other results, with some large entries: if $d = 967$,

$$33205563567547^3 + (-33201628358236)^3 = 967 \cdot 237872527101^3;$$

and—oh yes, some very much larger, as one with an x having 51 digits (when $d = 2677$).*

For $x^3 + y^3 = dz^3$, we know some d for which there are only a finite number of solutions, like $d = 3$. Others ($d = 13$, for example, or 16) splendidly produce an infinite number. Finding such d isn't actually hard, but it isn't the real issue; finding the structure of the solutions for a specific d is. Think of the complexity of the solutions we know so well for $x^2 + y^2 = z^2$, with its fan of practical applications and theoretical implications. Will there be similar riches for each of the d in our cubic equation? Will each star of a d in this $x^3 + y^3 = dz^3$ nebula have its own idiosyncratic planetary system? What is the way in from this question—and is it a way into a jungle or an overgrown city of observatories?

Let's narrow down our looking a second time to a more manageable vista, and work from now on with some specific d. You'll find that we stand at the entrance to an invitingly green tunnel.

You remember from Chapter Seven that the modern way to find all whole numbers x, y, z satisfying $x^2 + y^2 = z^2$ was to recast them as rational points x/z, y/z on a circle of radius 1, and then let a secant line pinned on this circle at $(-1, 0)$ sweep through it, reading off intersections with the circle when its slope was rational (for then these intersections would have rational coordinates too). The hidden power of this approach is that it generalizes from circles to other curves on the plane—such as our $x^3 + y^3 = dz^3$, when it too is rewritten as $\left(\dfrac{x}{z}\right)^3 + \left(\dfrac{y}{z}\right)^3 = d$.

---

* In every case, these triples are ancestral, as were the Pythagorean triples we found: each begets an infinite family of multiples that work too.

Once again we would let a straight line anchored on it, at some point with rational coordinates, sweep its contours, arresting it when the line's slope is rational. Unfortunately, in this new context, that won't guarantee that the new point of intersection will also be "good"—i.e., have rational coordinates. It would, however, had we drawn our line through *two* good points, for then the line itself will have rational coordinates, and will intersect the curve in a third good point; this leads to more secant lines and more good intersections, in a nicely reciprocal way, either forever, or until the process eventually closes up in a finite circuit.

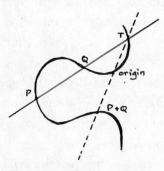

The point-and-secant technique on a generic cubic curve.

What is the minimal number of these 'starter points' needed, for a particular *d*, to generate all the rest? Thanks to Louis Mordell—nicknamed 'X, Y, Z' by his schoolmates—we've known, since 1922, that this minimal number will always be finite—but will it be large for some *d* and small for others? Will they even all lie below a uniform upper bound? We don't know, we don't know.[1]

What's alluring in all of this ignorance are the merest whiffs of intriguing structure drifting through, here and there, suggesting patterns of profound order: gold threads woven into a cushion tucked in the howdah on the back of the elephant that many a mathematician is groping over in the dark.

THE ABC CONJECTURE

What a universe there is of equations with three variables a, b, and c, standing for integers, and raised in different ways to powers. Call this universe Diophantine, after that astonishing explorer, Diophantus, whom you met briefly in Chapter Seven.

We know how to find the infinitely many solutions of $a^n + b^n = c^n$ when $n=2$, and Fermat tells us that there are no solutions in natural numbers when $n>2$.

The later Fermat-Catalan Conjecture claims that only finitely many solutions exist for $a^m + b^n = c^k$, (where all of these letters stand for natural numbers greater than 1; a, b, and c are relatively prime, and m, n, and k obey the further constriction that $(1/m) + (1/n) + (1/k) < 1$). Ten solutions—such as $2^5 + 7^2 = 3^4$—have been found so far, but the conjecture has yet to be proved.

The Dutch number theorist Robert Tijdeman suggested in the 1970s that there are also only finitely many solutions for $a^m + d = b^n$, for a fixed integer d greater than 1, with a, b, m, and n integers greater than 1. This too remains unproven.

And a page or two ago we flew past the spiral nebula $a^3 + b^3 = dc^3$, seeing many solutions when $d=7$, say, and knowing there are infinitely many when $d=13$ or 16, for example.

Mostly, however, this Diophantine universe is a mystery to us. If we can't yet tell its shape and the forces holding it together, could we at least find out in general how to distinguish cases with finitely many solutions from those with infinitely many? Or at any rate come up with a sufficient condition for the former—a condition that would say: if such and such is true, then there will be only a finite number of solutions?

Such a condition was proposed in 1985 by Joseph Osterlé and David Masser: the ABC Conjecture. To put it in a slightly weakened form: there will be only finitely many solutions when $a + b = c$, all are relatively prime positive integers, and c is greater than r^2, where r is the product of the primes dividing a, b, and c. Even in this weakened form, Osterlé and Masser's conjecture is at present out of our reach.

It took 358 years to settle Fermat's Last Theorem, but it has no hinterland: no structural consequences. Why should the ABC Conjecture seem to lead, past the proofs of these conjectures, to a deep understanding of our Diophantine universe?

You wouldn't be surprised were such a dynamo almost impenetrably packaged, like your car's control module—or, for that matter, like the brain in your skull. To grasp what effects the conjecture would have, were it proven, would be to bring yourself up-to-date in modern number theory.

And as for whether there really *is* a mechanism inside the housing, and it works, and its gears can be meshed with those outside it in a proof—that seems a prospect more like decades than years away. As you read this, a grid computing system in Holland is churning out triples (a, b, c) with $c > r^2$, as its programmers scan the output for patterns; and people around the world are teasing at and tinkering with the machinery that drives this condition—which is anyway only sufficient, not necessary, to distinguish merely between finitely and infinitely many solutions to the cluster of problems in this universe. We aren't even yet, it seems, in the antechamber to the presence room outside the hall where the profound answers are enthroned.

OUTWARD FROM SAMOS

Oh, and how many right triangles have rational sides and integral area? And can we find all non-Euclidean Pythagorean triples? And . . .

When does no more than a puzzle become a profundity, with the pieces all locked together in this Chinese box of a world? When it becomes structural.

And is Pythagoras not endlessly reborn past the thousand proofs, in the generalizations and conjectures of what passes for his theorem?

The world as it yet might be, unguessed, cradles the world as it only is.

Reaching Through—
or Past—History?

"I often think it odd," says Catherine Morland in *Northanger Abbey*, "that history should be so dull, for a great deal of it must be invention." The philosopher J. L. Austin (who always hoped his books would be mistaken for Jane's) couldn't have sparked a livelier debate. What should history be—objective or subjective? Could it be either, or both? These are especially sharp alternatives in our little corner of it, where we have so much to do with proofs—which are ligatures in time to eternal truths. Yet the proofs are the work of provers, and the words of the leading Pythagorean scholar, Walter Burkert, are a tinnitus in the ear: "One is tempted to say that there is not a single detail in the life of Pythagoras that stands uncontradicted."[1]

Perhaps there are no facts, as a sparring partner of Austin's once claimed. Ours, at least, are very far away. Theory crusts them over. Intention generates names and dates, and intentions are puffball.

Looking back, we lose sight too of those minute, informing leanings, so negligently known at the time, that ended up prodding balanced alternatives in the direction they ultimately took. We replace them, in our architectural zeal, with abstract forces grandly materializing in events: the Being that brackets all Becoming. This Being, however, has noticeably contingent qualities. We still cast it largely in a linear, causal mode rather than that of a network. You never hear anyone ask, "What is *a* meaning of things?"

In the Freer Gallery of Art in Washington there hangs a magnificent *Nocturne* of Whistler's, showing Battersea Reach. There is the bridge, and there a barge beneath. Those ripples of yellow are riding lights; the darker verticals are masts and, beyond, stacks on the bank. Or is it a quartering of the canvas into water, fog, and moonlight—or rather, into lighter blues shading to darker: not so much representation as composition? Expert knowledge of time, place, and biography might reveal further details of the shore—or the brushwork—or the intentions. Past a certain threshold of analysis, however, we would come up with only ground ultramarine and canvas threads. To press too far beyond an intriguing scatter of data is to study not history but historiography.

Historians need an inner justicer to arbitrate among their capacities: an eye for details, an imagination to vivify, and a mind to deduce the likely to and from them, with an engaged spirit relishing presence and seeking out significance. We easily get lost in these contrary enterprises, taking our singular selves as the measure of all things, turning into unwitting partisans of those we study, recasting a simplifying hypothesis as world order. And there are all the little treasons of history's clerics, like reading the prior generation's orthodoxies as heresy. This was once thought Oedipal but is no more than opportunistic, in the pursuit not of history but of the historian's profession.

With our scanty evidence so corrupt and corruptible, what should we do? In what tense should our stories be written—the historical past, whose every ending acknowledges its fictional status? In what mood—the subjunctive, where "must" is no more than projected "ought"? Or aim to establish as probable a context as we can shore up in a sea of the possible, so that within it we may measure in common with distant others the depth of a startled insight, for all the incommensurability of our lives?

Yet there is the outlook above, starring each of our skies with theorems and their awesome premonitions of order. Let us indeed eke out the lives of the provers with our imaginations, and reckon the past from the bearings this inner compass gives: but our course is set by the uncanniness of impersonal truths. "He whose oracle is in Delphi," said Heraclitus, "neither affirms nor denies, but indicates." And Pythagoras, bending over the tripod of the theorem that bears his name, no more than points toward the abstract, which informs our every singular.

Acknowledgments

We've been very lucky in the inspiring help we've had, with so many different aspects of this book, from Barry Mazur, Amanda and Dean Serenevy, John Stillwell, Jon Tannenhauser, and Jim Tanton. Eric Simonoff has been, as ever, a stimulating and benign presence. The *OED* speaks of an editor as one who gives to the world. Peter Ginna exemplifies this: the book was his idea, and his patience, humor, and encouragement have brought it out.

Our immense thanks to friends, acquaintances, and strangers who have variously set us straight or pointed out a path where we saw none: David Domotor, Grant Franks, Gene Golovchenko, Jean Jaworski, Ulla Kasten, Oliver Knill, Jim Propp, Jan Seymour-Ford, and Michael Zaletel.

Notes

1. In Aldous Huxley, *Little Mexican* (London: Chatto & Windus, 1924).
2. Heath, *A History of Greek Mathematics* I.142. Heath notes the anachronism: Thales precedes Pythagoras. On the other hand, it isn't Pythagoras whom Callimachus names but the Homeric hero Euphorbus, whom Pythagoras claimed to have reincarnated, and who did indeed precede Thales.
3. Ibid., I.121, citing Herodotus II.109.
4. Ibid. Heath points out that this claim occurs in Greek historians, who may all be no more than elaborating on the bare statement, given above, from Herodotus.
5. Wikipedia, "Berlin Papyrus", http://en.wikipedia.org/wiki/Berlin_Papyrus.
6. The passage on the stone circle in Egypt is from http://www.philipcoppens .com/carnac.html, "Counting Stones". On the analysis of Thom's data, see M. Beech, "Megalithic Triangles", *Journal of Recreational Mathematics* 20, no. 3 (1988).

CHAPTER 2

1. The passage is from Samuel Noah Kramer, "Schooldays: A Sumerian Composition Relating to the Education of a Scribe", *Journal of the American Oriental Society* 69, no. 4 (October–December 1949): 208.
2. The outline of our summary is based largely on Høyrup, "Mesopotamian Mathematics".
3. Friberg, *A Remarkable Collection*, 434fn.
4. Robson, "Words and Pictures", 111–112, and her "Three Old Babylonian Methods", 70.
5. Ramanujan to Hardy, letter of February 27, 1913: "If I tell you this [namely, that the sum of all the natural numbers equals $-1/12$], you will at once point out to me the lunatic asylum as my goal."
6. Friberg, *Amazing Traces*, 36.

7. Herodotus II.109, Diogenes Laertius II.1–2, Suda, s.v., cited by G. S. Kirk, J. E. Raven, and M. Schofield, 100–103.

8. Aristotle, *Physics* G4, 203a10.

9. As listed in Friberg, *Remarkable Collection*, 449: "Damerow listed all mathematical cuneiform texts known to him, in which the theorem is used either directly or indirectly. . . . Høyrup wrote a brief note with a discussion of nine [further] examples. . . . The Høyrup/Damerow list is updated below, and made considerably more complete and explicit." We have confined ourselves to Old Babylonian texts only, since examples from the Late Babylonian/Seleucid might well have been back-influenced by developments in Greece.

10. Robson, "Neither Sherlock Holmes nor Babylon", 185.

11. Jens Høyrup, "Changing Trends", 23–24fn.

12. Robson, "Neither Sherlock Holmes", 185.

13. Høyrup, "Changing Trends", 24.

14. David E. Joyce, "Plimpton 322", http://aleph0.clarku.edu/~djoyce/mathhist/plimpnote.html (1995).

15. Robson, "Neither Sherlock Holmes", 182–83; Jens Høyrup, "Mesopotamian Mathematics", in *Cambridge History of Science* (Cambridge: Cambridge University Press) I.7.

16. R. Creighton Buck, "Sherlock Holmes in Babylon", *American Mathematical Monthly*, May 1980, 335–45.

17. Friberg, *Remarkable Collection*, 337, YBC 6967.

18. Robson, "Neither Sherlock Holmes", 167–206.

19. *Meno* 82b–85b.

20. See, for example, Høyrup, "Mesopotamian Mathematics", 10.

21. VAT 6598 and BM 96957. See Robson, "Three Old Babylonian Methods", 51–72.

22. Ibid., 63, 64.

23. Friberg, *Amazing Traces*, 395.

24. P. Bergh, in *Zeitschrift für Math. u. Physik* xxx. Hist-Litt. Abt., 135, cited in Heath's *Euclid* I.400–401.

25. Heath, *A History of Greek Mathematics* I.61–63.

26. Robson, "Three Old Babylonian Methods", 66.

27. This is also the value reported by Ptolemy for $\sqrt{2}$ in his *Syntaxis* I.10, ed. Heiberg I.32.10–35.16.

CHAPTER 3

1. While people speak loosely of a tradition ascribing this proof to Pythagoras, in fact the earliest record we have of it dates from about the second century B.C. in China—see our discussion of it in Chapter Five.

2. See his *Euclid* I.354–55 and his *Greek Mathematics* I.149.

3. Burkert, 110–12.

4. Aristotle, cited in G. S. Kirk, J. E. Raven, and M. Schofield, 228.

5. Burkert, 140 n. 109.

6. Pythagoras's previous incarnations: ibid., 138, 140 n. 110.

7. Acusmata listed in ibid., 166–92.

8. Ibid., 179.

9. Ibid., 115–17.

10. Dodgson, *A New Theory of Parallels* (London: Macmillan, 1888), 16.

11. Burkert, 156.

12. Ibid., 142; Kirk, Raven, and Schofield, 228.

13. Biting the serpent, stroking the eagle, and advent of the white bear: Burkert, 142–43.

14. Ibid., 433.

15. Bertrand Russell, *Autobiography* (New York: Simon & Schuster, 1929), 3:330.

16. Cited in Barry Mazur's "How Did Theaetetus Prove His Theorem?" in *The Envisioned Life: Essays in Honor of Eva Brann*, ed. P. Kalkavage and E. Salem (Philadelphia: Paul Dry Books, 2007), 227.

17. Music of the spheres, and hearing it from birth: Aristotle, *De Caelo* B9, 290b12.

18. Philolaus, from fragment 6, Stobaeus *Arith.* I.21.7d, quoted in Kirk, Raven, and Schofield, 327.

19. This non-Pythagorean view is associated with Aristoxenus. See Burkert, 369 ff.

20. On whether the inner and outer Pythagorean crises coincided, see ibid., 207.

21. The view that the philosophy of Parmenides exerted a decisive influence on Pythagorean and Greek mathematical thought is forcefully put by Szabó, 248–57 and passim. See too Burkert, 425.

22. The octave, for example, could not be divided in half. The question of whether this discovery led to or followed from its mathematical analogue is much debated. Burkert, 370 n. 4, thinks it unlikely; Szabó, 173–74, 199 ff., highly probable.

CHAPTER 4

1. It is called on directly or indirectly in the proofs of the last six propositions of Book II, five more in Book III (14, 15, 35, 36, 37), and three in Book IV (10, 11, 12).
2. Schopenhauer's names for I.47 are quoted in Heath's *Euclid* I.354.
3. The etymology is Heath's (in his *Euclid* I.418) by way of Skeat.
4. Quoted in ibid., I.350.
5. Euclid's three lost books of Porisms, ibid., I.10 ff.
6. Our thanks to our friend and Internet virtuoso, Jon Tannenhauser, for retrieving the magazine's cover from the World Wide Web.
7. Heath, *Euclid* I.417–18.
8. Quoted by Richard K. Guy in "The Lighthouse Theorem", *Mathematical Association of America Monthly*, February 2007, 124.

CHAPTER 5

1. George MacDonald Fraser, *Quartered Safe Out Here* (London: Harvill Press, 1992), 150.
2. We encountered this proof in Barry Mazur, "Plus symétrique que la sphère", *Pour la Science* no. 41, October–December 2003, "Mathématiques de la sphère", 4.
3. See Karine Chemla, "Geometrical Figures and Generality in Ancient China and Beyond: Liu Hui and Zhao Shuang, Plato and Thabit ibn Qurra", *Science in Context* 18, no. 1 (2005): 159.
4. We are indebted for our information about Emma Coolidge Weston not only to Jan Seymour-Ford but to Jean Jaworski, executive administrator of the New Hampshire Association for the Blind.
5. All quotations in this paragraph are from Allyn Jackson, "The World of Blind Mathematicians", *Notices of the American Mathematical Society* 59, no. 10, November 2002, 1246–51.
6. The earlier Bhaskara lived around A.D. 629 in Kerala, the later—often called Bhaskara the teacher—not only five centuries but twelve hundred miles away, at the astronomical observatory in Ujjain.
7. This material is from Kim Plofker's "Mathematics in India", in Katz, 402–3, 411, 477.
8. Plato, *Meno* 77A.
9. Wagner, 71–73.
10. D. G. Rogers, "Putting Pythagoras in the Frame", *Mathematics Today* 44 (June 2008): 123–25. Rogers speaks of his reconstruction as "only a

matter of mathematical *play*, without suggestion that this has any *historical* basis."

11. Chemla, op. cit., 127 n. 11.

12. This well-known diagram is reproduced from ibid., 149. We follow Chemla's interpretation.

13. Quoted in Paul Arthur Schilpp, *Albert Einstein: Philosopher-Scientist* (New York: Tudor), 1951, 342.

14. Chemla, op. cit., 141.

15. Elsewhere he certainly spared no pains to prove (by inscribed and circumscribed polygons of up to 192 sides) that π isn't 3 but somewhat less than 3.142704, and he says: "Yet a tradition [that $\pi = 3$] has been passed down from generation to generation and no one cares to check it. So many scholars followed the tradition that their error has persisted. It is hard to accept without a convincing demonstration" (which he then gives). Of course this correcting (as he says, of fellow scholars, rather than of the venerable sages) amounts to redefining a value through repeated calculation—very different in spirit from contemplating the role and authority of the 3-4-5 right triangle as exemplar rather than example. The passage from Liu Hui is from Joseph W. Dauben, "Chinese Mathematics", in Victor Katz (op. cit.), 235, and Liu Hui's demonstration is on 236–37 there.

16. We have seen this proof variously attributed to Frank Burk (*College Mathematics Journal* 27, no. 5 [November 1996]: 409) and, on http://www .cut-the-knot.org/pythagoras, to Geoffrey Margrave of Lucent Technologies (#41 on this site), with one variation by "James F." and another referred to G. D. Birkhoff and R. Beatley, *Basic Geometry* (a 2000 republication of a text that first appeared in 1959), 92.

17. Michael Hardy, "Pythagoras Made Difficult", *Mathematical Intelligencer* 10, no. 3 (1988): 31. See also http://www.cut-the-knot.org/pythagoras, proof #40.

18. See Rüdiger Thiele, "Hilbert's Twenty-Fourth Problem", *Mathematical Association of America Monthly*, January 2003, 1–24.

19. "Proof of the Theorem of Pythagoras", Alvin Knoer [*sic*], *Mathematics Teacher* 18, no. 8 (December 1925): 496–97.

20. P. W. Bridgman, *Dimensional Analysis* (New Haven: Yale University Press, 1922). Thanks to Jene Golovchenko in Harvard's physics department, we've been able to trace this application of dimensional analysis to the Pythagorean Theorem back to A. B. Migdal's *Qualitative Methods in Quantum Theory* (1977), although G. Polya's discussion on pp. 16–17 of

his 1954 *Induction and Analogy in Mathematics* could be seen as anticipating it. Perhaps its origin lies farther back in time.

21. This proof of Tadashi Tokeida's is from his "Mechanical Ideas in Geometry", *American Mathematical Monthly*, October 1998, 697–703. His love of children's books will make him sympathize with our vicarious Dorothy. See Scott Brodie's comments on this proof at http://www.cut-the-knot .org, "'Extra-geometric' Proofs of the Pythagorean Theorem". See too four other physics-inspired proofs in Mark Levi's *The Mathematical Mechanic* (Princeton: Princeton University Press, 2009): two via the equilibrium of a sliding ring, attached by springs to the ends of its track; another pair involving the kinetics of skating, with a variation picturing two equal hurled masses flying apart at right angles to their trajectory. Levi notes the contrast between the rigor of the mathematical proof and the conceptual character of the physical.

22. We found this tiling of the torus and its proof in Ian Stewart's "Squaring the Square", *Scientific American*, July 1997, 96.

23. We now have a second record of the thinking that led to a proof of our theorem. The ingenious Rory O'Brien, in Ireland, sent us this:

"Mulling over your account of Alvin Knoerr's lovely ingenuity, I experimented with number-line constructions modelling multiplication and division. I tried to get the constructions to model, and thus verify the equivalence of, each side of Alvin's step 14, which would then lead directly to Pythagoras's proposition. I suppose this approach succeeded in some sense—I now found myself with a version of an old proof by similar triangles, but an unnecessarily elaborate version that took the scenic route! Some of the structural superfluity of my efforts seemed to have a nice geometrical symmetry about it, so I focused on this as a possible basis for a fresh approach. This symmetry is constructed very simply but exhibits lots of equalities and proportions. I began by writing down as many of these as I could spot. I then spent a whole night torturing these truths algebraically (not that truth seems to mind—it knows what it is). The aim was to see if the two sides of equation $AB^2 = AC^2 + BC^2$ could be expressed entirely in terms common to both, in the hope of thus sussing their elusive equivalence. At 3:00 a.m. lucky things began to happen. At 5:30 a.m. I noticed that I was in possession of a secure avenue to sufficient granularity."

And here is Rory's proof:

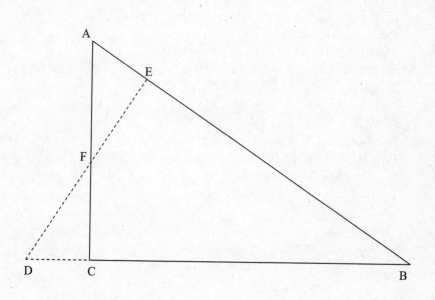

Let ABC be a right-angled triangle with hypotenuse AB.

AB is therefore the longest side.

Extend BC to D so that BD = AB.

Cut AB at E so that BE = BC.

Then D and E must lie on opposite sides of AC.

Therefore AC and DE cannot be parallel and cannot coincide: they
    meet at a point F.

Now triangles ABC and BDE have an angle in common at B.

Angle B is enclosed by sides BD = AB and sides BE = BC.

Therefore triangle ABC = BDE.

Therefore angle D = angle A, and angle E = angle C = right angle.

Therefore triangles AEF and CDF have equal angles at A and D.

But AEF and CDF also have a right angle, at E and C respectively.

Then AEF and CDF also have equal (and opposite) angles at F too.

By construction AE = CD.

Therefore triangle AEF = CDF.

Now all four triangles, ABC, BDE, AEF, CDF have a right angle and
    also angle A or D.

But angle A = D, so the remaining angles of all four triangles must also be equal.

Therefore the sides of the four triangles are in proportion.

Then AF ÷ AE = AB ÷ AC

Hence (AC – CF) ÷ AE = (AE + BE) ÷ AC

Hence (AC – EF) ÷ AE = (AE + BC) ÷ AC

By cross multiplication, $AC^2 - AC.EF = AE^2 + AE.BC$

But by another proportion of sides, AC ÷ AE = BC ÷ EF

So AC.EF = AE.BC

Therefore $AC^2 - AE.BC = AE^2 + AE.BC$

Then $AC^2 = AE^2 + 2AE.BC$

Now AB = BE + AE = BC + AE

Therefore $AB^2 = (BC + AE)(BC + AE)$

$= BC^2 + AE^2 + 2AE.BC$

$= BC^2 + AC^2$

—Rory O'Brien, 15 December 2011

CHAPTER 6

1. Since Apollonius would have known the law of cosines, Stewart's proof is occasionally backdated to him by fervent Pergaphiles.

2. Dijkstra's generalization of the Pythagorean Theorem is in his manuscript EDW975, http://www.cs.utexas.edu/users/EWD/transcriptions/ EWD09xx/EWD975.html. His fantasy about Mathematics Inc. is described in the Wikipedia article on him, "Edsger Wybe Dijkstra". The varying opinions of him, and contradictory anecdotes, appear in (among others) Krzystof Apt, "Edsger Wybe Dijkstra: A Portrait of a Genius", http://homepages.cwi.nl/~apt/ps/dijkstra.pdf); Mario Szegedy, "In Memoriam Edsger Wybe Dijkstra (1930–2002)", http://www.cs.rutgers.edu/ ~szegedy/dijkstra.html; and J. Strother Moore, "Opening Remarks. . . .", www.cs.utexas.edu/users/EWD/memorial/moore.html. Nano-Dijkstras are referred to in Alan Kay's 1997 address to OOPSLA, on YouTube.

3. Leonhard Euler, *Introduction to Analysis of the Infinite*, Book I, Chapter VIII, 132.

4. Most of the biographical material on Johannes Faulhaber is from Matthias Ehmann, "Pythagoras und Kein Ende", http://did.mat.uni-bayreuth .de/~matthias/geometrieids/pythagoras/html/node11.html. Descartes's ideas about the orthogonal tetrahedron in four dimensions are in his

Cogitationes Privatae X, 246–48. For Tinseau, see http://www-history.mcs
.st-andrews.ac.uk/Biographies/Tinseau.html. The fullest life of de Gua
that we found is on http://fr.wikipedia.org/wiki/Jean-Paul_de_Gua_de_
Malves.

5. This proof is due to Jean-P. Quadrat, Jean B. Laserre, and Jean-B. Hiriart-
Urruty, "Pythagoras' Theorem for Areas", *American Mathematical Monthly*,
June–July 2001, 549–51. An alternative proof is in Melvin Fitting,
"Pythagoras' Theorem for Areas—Revisited", http://comet.lehman.cuny
.edu/fitting/bookspapers/pythagoras/pythagoras.pdf.

CHAPTER 7

1. The method of *diaresis*, or division, as exemplified in Plato's dialogue *The Sophist*.
2. Herbert Spencer Lecture, "On the Methods of Theoretical Physics",
Oxford University, June 10, 1933.
3. The long but beautiful story behind this insight is well explained in Joseph
Silverman's *A Friendly Introduction to Number Theory* (Upper Saddle River,
N.J.: Prentice Hall, 1997), 169.

CHAPTER 8

1. "Et tout ce que l'Idylle a de plus enfantin." Baudelaire, *Paysage*.
2. Henry Reed, "Judging Distances", which first appeared in *New Statesman
and Nation* 25, no. 628 (March 6, 1943): 155. Published in *Collected Poems*,
ed. John Stallworthy (Oxford University Press, 1991).
3. This proof is modified from Klaus Lagally's, as reported in D. J. Newman,
"Another cheerful fact about the square of the hypotenuse", *Mathematical
Intelligencer* 15, no. 2 (1993): 58.
4. From George Starbuck's poem "The Unhurried Traveler in Boston".
5. And we, the wise men and poets
　　Custodians of truths and of secrets
　Will bear off our torches and knowledge
　　To catacombs, caverns and deserts.

The mathematician L. A. Lyusternik, former member of Lusitania, quoted
in Graham and Kantor, 101.
6. Quoted in Abe Shenitzer and John Stillwell, eds., *Mathematical Evolutions*.
(Washington: the Mathematical Association of America, 2002), 44–45.
7. Robert Browning, "By the Fire-Side."

CHAPTER 9

1. From John Aubrey's *Brief Lives II*, 220–21.
2. These include John Q. Jordan and John M. O'Malley Jr., "An Implication of the Pythagorean Theorem", *Mathematics Magazine* 43, no. 4 (September 1970), 186–89; Jingcheng Tong, "The Pythagorean theorem and the Euclidean parallel postulate", *International Journal of Mathematical Education in Science and Technology* 32 (2001): 305–8; and David E. Dobbs, "A single instance of the Pythagorean theorem implies the parallel postulate", *International Journal of Mathematical Education in Science and Technology* 33, no. 4 (July 2002): 596–600.
3. This sequence of lemmas and theorems is from Scott E. Brodie, "The Pythagorean Theorem Is Equivalent to the Parallel Postulate", http://www.cut-the-knot.org/triangle/pythpar/PTimpliesPP.shtml.
4. See, for example, what Plato has Simmias say in the *Phaedo*, 86B.
5. Our thanks to Jim Tanton for working out this elegant proof.

CHAPTER 10

1. You will find much of what we do know in Noam Elkies's and Nicholas Rogers's "Elliptic Curves $x^3 + y^3 = k$ of High Rank", in *Algorithmic Number Theory* 3076 (Berlin: Springer, 2004): 184–93; Elkies, "Tables of fundamental solutions (x, y, z) of $x^3 + y^3 = pz^3$ with p a prime congruent to 4 mod 9 and less than 5000 or congruent to 7 mod 9 and less than 10000", http://www.math.harvard.edu/~elkies/sel_p.html; and Elkies, "Tables of fundamental solutions (x, y, z) of $x^3 + y^3 = p^2z^3$ with p a prime congruent to 4 mod 9 and less than 1000 or congruent to 7 mod 9 and less than 666", http://www.math.harvard.edu~elkies/sel_p2.html.

AFTERWORD

1. Burkert, 109.

Selected Bibliography

Beauregard, R. A., and E. R. Suryanarayan. "Arithmetic Triangles." *Mathematics Magazine* 70, no. 2 (April 1997): 105–15.

———. "Proof Without Words: Parametric Representation of Primitive Pythagorean Triangles." *Mathematics Magazine* 69, no. 3 (June 1996): 189.

Borzacchini, Luigi, David Fowler, and David Reed. "Music and Incommensurability." E-mail exchange on http://mathforum.org/kb/thread.jspa?threadID=384376&messageID=1186550, July 1999.

Bogomolny, Alexander. "Pythagorean Theorem." http://www.cut-the-knot.org/pythagoras/index.shtml.

Burkert, Walter. *Lore and Science in Ancient Pythagoreanism.* Trans. Edwin L. Minar Jr. Cambridge: Harvard University Press, 1972.

Fowler, D. H. *The Mathematics of Plato's Academy: A New Reconstruction.* New York: Oxford University Press, 1987.

Fowler, David, and Eleanor Robson. "Square Root Approximations in Old Babylonian Mathematics: YBC 7289 in Context." *Historia Mathematica* 25 (1998): 366–78.

Friberg, Jöran. *Amazing Traces of a Babylonian Origin in Greek Mathematics.* Hackensack, N.J.: World Scientific, 2007.

———. *A Remarkable Collection of Babylonian Mathematical Texts.* New York: Springer, 2007.

————. *Unexpected Links Between Egyptian and Babylonian Mathematics.* Hackensack, N.J.: World Scientific, 2005.

Graham, Loren, and Jean-Michel Kantor. *Naming Infinity.* Cambridge: Belknap Press of Harvard University Press, 2009.

Grattan-Guinness, I., ed. *Companion Encyclopedia of the History and Philosophy of the Mathematical Sciences.* London and New York: Routledge, 1994.

Heath, Thomas. *A History of Greek Mathematics.* 2 vols. 1921; rpt. New York: Dover, 1981.

————, trans. and ed. *The Thirteen Books of Euclid's Elements.* 2nd ed. 3 vols. 1925; rpt. New York: Dover, 1956.

Høyrup, Jens. "Mesopotamian Mathematics." http://akira.ruc.dk/~jensh/ Selected%20themes/Mesopotamian%20mathematics/index.htm. D:11

————. "Was Babylonian Mathematics Created by 'Babylonian Mathematicians'?" Paper presented to 4. Internationales Kolloquium des Deutschen Orient-Gesellschaft "Wissenskultur im Alten Orient," Münster/Westf., February 20–22, 2002, preliminary version, February 14. http://akira.ruc.dk/ ~jensh/Selected%20themes/Mesopotamian%20mathematics/index.htm. G:21

————. "Changing Trends in the Historiography of Mesopotamian Mathematics—An Insider's View." Revised Contribution to the Conference "Contemporary Trends in the Historiography of Science," Corfu, May 27–June 1, 1991. http://akira.ruc.dk/~jensh/Selected%20themes/Mesopota mian%20mathematics/index.htm. B:35

Kahn, Charles H. *Pythagoras and the Pythagoreans.* Indianapolis: Hackett, 2001.

Katz, Victor, ed. *The Mathematics of Egypt, Mesopotamia, China, India, and Islam.* Princeton: Princeton University Press, 2007.

Kirk, G. S., J. E. Raven, and M. Schofield. *The Presocratic Philosophers.* 2nd ed. Cambridge: Cambridge University Press, 1984.

Kutrovátz, Gábor. "Philosophical Origins in Mathematics? Árpád Szabó Revisited", 13th Novembertagung on the History of Mathematics, Frankfurt, October 31–November 3, 2002.

Loomis, Elisha S. *The Pythagorean Proposition*. 2nd ed. N.p.: Masters and Wardens Association of the 22nd Masonic District of the Most Worshipful Grand Lodge of Free and Accepted Masons of Ohio, 1940.

Loy, Jim. "The Pythagorean Theorem." http://www.jimloy.com/geometry/pythag.htm.

Maor, Eli. *The Pythagorean Theorem: A 4,000-Year History*. Princeton: Princeton University Press, 2007.

Mazur, Barry. "Plus symétrique que la sphere." *Pour la Science* no. 41, "Mathématiques de la sphère," October–December 2003.

Nelsen, Roger B. *Proofs Without Words*. Washington, D.C.: Mathematical Association of America, 1993.

———. *Proofs Without Words II*. Washington, D.C.: Mathematical Association of America, 2000.

Netz, Reviel. *The Shaping of Deduction in Greek Mathematics*. Cambridge: Cambridge University Press, 1999.

Neugebauer, O. *The Exact Sciences in Antiquity*. Princeton: Princeton University Press, 1952.

Newman, D. J. "Another Cheerful Fact About the Square of the Hypotenuse," and Klaus Lagally, "Solution." *Mathematical Intelligencer* 15, no. 2 (1993): 58.

Pont, Graham. "Philosophy and Science of Music in Ancient Greece: The Predecessors of Pythagoras and Their Contribution." *Nexus Network Journal* 6, no. 1 (Spring 2004). http://www.emis.de/journals/NNJ/index.html.

Porter, Gerald J. "k-volume in R^n and the Generalized Pythagorean Theorem." *Mathematical Association of America Monthly*, March 1996, 252–56.

Riedweg, Christoph. *Pythagoras: His Life, Teaching, and Influence*. Ithaca: Cornell University Press, 2005.

Robson, Eleanor. "Mathematics, Metrology, and Professional Numeracy." In *The Babylonian World*, ed. Gwendolyn Leick. London: Routledge, 2007.

———. "Neither Sherlock Holmes nor Babylon: A Reassessment of Plimpton 322." *Historia Mathematica* 28 (2001): 167–206.

———. "Three Old Babylonian Methods for Dealing with 'Pythagorean' Triangles." *Journal of Cuneiform Studies* 49 (1997): 51–72.

———. "Words and Pictures: New Light on Plimpton 322." *Mathematical Association of America Monthly*, February 2002, 105–20.

Shanks, Daniel. *Solved and Unsolved Problems in Number Theory*. Washington, D.C.: Spartan Books, 1962.

Sierpinski, Waclaw. *Pythagorean Triangles*. New York: Graduate School of Science, Yeshiva University, 1962.

Staring, Mike. "The Pythagorean Proposition: A Proof by Means of Calculus." *Mathematics Magazine* 69, no. 1 (February 1996): 45–49.

Szabó, Árpád. *The Beginnings of Greek Mathematics*. Trans. A. M. Ungar. Dordrecht, Boston, and London: Reidel, 1978.

von Fritz, Kurt. "The Discovery of Incommensurability by Hippasus of Metapontum." *Annals of Mathematics*, 2nd Ser., 46, no. 2 (April 1945): 242–64.

Weisstein, Eric W. "Pythagorean Theorem." From *MathWorld*—a Wolfram Web Resource. http://mathworld.wolfram.com/PythagoreanTheorem.html.

Wagner, Donald B. "A Proof of the Pythagorean Theorem by Liu Hui (third century A.D.)." *Historia Mathematica* 12 (1985): 71–73.

Index